THE LIFE, LOVES & INVENTIONS OF

ANTONIO MEUCCI

FATHER OF THE TELEPHONE

HOLD THE LINE

Francesca Valente

THE MENTORIS PROJECT

Mentoris Project
745 South Sierra Madre Drive
San Marino, CA 91108

Cover photo: Emilio Ereza / Alamy Stock Photo

Cover design: Suzanne Turpin

More information at www.mentorisproject.org

ISBN: 978-1-947431-59-1

Library of Congress Control Number: 2025949016

All net proceeds from the sale of this book will be donated to the Mentoris Project whose mission is to support educational initiatives that foster an appreciation of history and culture to encourage and inspire young people to create a stronger future.

Publisher's Cataloging-In-Publication Data (Provided by Cassidy Cataloguing Services, Inc.)
Names: Valente, Francesca, author.
Title: Hold the line : the life, loves & inventions of Antonio Meucci, father of the telephone / Francesca Valente.
Other titles: Life, loves & inventions of Antonio Meucci, father of the telephone
Description: San Marino, CA : The Mentoris Project, [2025] | Includes bibliographical references.
Identifiers: ISBN: 9781947431591
Subjects: LCSH: Meucci, Antonio, 1808-1889. | Inventors--Italy--Biography. | Telephone--History. | LCGFT: Biographies. | BISAC: BIOGRAPHY & AUTOBIOGRAPHY / Science & Technology. | BIOGRAPHY & AUTOBIOGRAPHY / General. | BIOGRAPHY & AUTOBIOGRAPHY / Historical.
Classification: LCC: TK6018.M4 V36 2025 | DDC: 609.2--dc23

The Mentoris Project is a series of novels and biographies about the lives of great men and women who have changed history through their contributions as scientists, inventors, explorers, thinkers, and creators. The Barbera Foundation sponsors this series in the hope that, like a mentor, each book will inspire the reader to discover how she or he can make a positive contribution to society.

Contents

Foreword

First and foremost, Mentor was a person. We tend to think of the word *mentor* as a noun (a mentor) or a verb (to mentor), but there is a very human dimension embedded in the term. Mentor appears in Homer's *Odyssey* as the old friend entrusted to care for Odysseus's household and his son Telemachus during the Trojan War. When years pass and Telemachus sets out to search for his missing father, the goddess Athena assumes the form of Mentor to accompany him. The human being welcomes a human form for counsel. From its very origins, becoming a mentor is a transcendent act; it carries with it something of the holy.

The Mentoris Project sets out on an Athena-like mission: We hope the books that form this series will be an inspiration to all those who are seekers, to those of the twenty-first century who are on their own odysseys, trying to find enduring principles that will guide them to a spiritual home. The stories that comprise the series are all deeply human. These books dramatize the lives of great men and women whose stories bridge the ancient and the modern, taking many forms, just as Athena did, but always holding up a light for those living today.

Whether in novel form or traditional biography, these books plumb the individual characters of our heroes' journeys. The power of storytelling has always been to envelop the reader in a vivid and continuous dream, and to forge a link with the subject. Our goal is for that link to guide the reader home with a new inspiration.

What is a mentor? A guide, a moral compass, an inspiration. A friend who points you toward true north. We hope that the Mentoris Project will become that friend, and it will help us all transcend our daily lives with something that can only be called holy.

—Robert J. Barbera, Founder, The Mentoris Project
—Ken LaZebnik, Founding Editor, The Mentoris Project

Introduction

The telephone is one of the most influential technological innovations in the field of communications. In the course of history, many individuals claimed to have invented it.

In Europe, Charles Bourseul, Innocenzo Manzetti, and Johann Philipp Reis made their cases; in the United States, Sylvanus D. Cushman, Amos Dolbear, Daniel Drawbaugh, Edward Farrar, James McDonough, Elisha Gray, Alexander Graham Bell, and Antonio Meucci all sought credit for it. It has been proven and now ruled that Meucci was indeed the first to invent the telephone, not Alexander Graham Bell, as most people still believe.

Meucci was a nineteenth-century Italian who was also fluent in French and Spanish. After moving to the U.S. in 1850, he devised a functioning electric telephone and used it daily. Looked down upon by American society in the way most European immigrants were at the time, and not articulate in English, Meucci was unable to defend his rights as an inventor. He always had to rely on an interpreter or a friend to convey the complex theories underlying his brilliant inventions, particularly in the last thirty-nine years of his life, spent in New York.

His nature was that of a dedicated Italian patriot and a compulsive, generous researcher. He put his intuitions and problem-solving abilities to serve society at large in order to improve people's quality of life. He never thought of his research in terms of mere moneymaking, but rather as a way to pursue personal growth and scientific progress. His discoveries were ahead of his time by at least thirty years.

When I was asked to write a new biography of Antonio Meucci and the invention of the telephone, I was afraid I was about to embark on a long, frustrating venture. I would have to examine one of the most corrupt periods in nineteenth-century America and the failings of the U.S. legal system, possibly without even coming close to understanding why it had rejected hundreds of challenges to A. G. Bell's claim to the telephone. I thought it would mean defying the authors of the many books that credited Bell with the telephone invention. The prospect seemed more than daunting.

In Bell's day, courts relied exclusively on pre-trial statements and depositions to establish the facts of a case. Nowadays the American legal system would require Bell to hand over his notebooks, including the incriminating evidence they contained. But these rules were not uniformly applied until 1938. Bell's correspondence and papers only became available after 1976, when they were donated by his family to the Library of Congress, and later digitized. Until then, they had safely remained for over a century in the family's possession, stored in a secluded room at the Washington, D.C., headquarters of National Geographic Society, whose founder and first president, Gardiner G.

Hubbard, was also the deus ex machina who, as you will soon find out, maneuvered A. G. Bell's patent in order to successfully present him as the only inventor of the telephone.

What I learned from my research into the invention of the telephone is that we must constantly challenge and interrogate even the most established of narratives, rather than "tacitly accepting the garbled story that is whispered from one generation to the next".[1]

—Francesca Valente

Chapter One

THE YEARS IN FLORENCE / FLORENTINE EDUCATION (1808–1835)

In Florence, which had recently fallen under Napoleon's rule, Antonio Meucci was born to Maria Domenica Pepi and Amatis Meucci on Wednesday, April 13, 1808, at 5 o'clock in the morning. He was the eldest of nine children, four of whom died in early childhood.

Amatis was thirty years old when Antonio was born. They lived in a house with three other families at via Chiara 475—now via dei Serragli 44—in the working-class district of San Frediano, on the left bank of the Arno river. Amatis had set aside enough wood to keep the fire going the whole night and guarantee plenty of hot water for the birth of his firstborn. He was so thrilled and proud when the midwife announced it was a boy and showed him the baby. Antonio weighed over 6.5 pounds, with black hair and a dark complexion just like his father's. Maria Domenica was lying in bed, exhausted and almost unable to speak. Amatis, on the other hand, was very excited and, with his wife's consent, decided to have Antonio baptized as soon as possible.

The following day he managed to have a carriage at his disposal for a few hours, and along with the baby, the midwife, and an aunt, crossed the Carraia bridge and reached the Baptistery of San Giovanni opposite the city's cathedral, Santa Maria del Fiore. Its octagonal marble-clad Romanesque baptistery, originally a Roman temple and later called by Dante "il bel San Giovanni," was in those years the place in Florence where infants, both rich and poor—including famous artists and poets such as Dante himself, and even members of the Medici family—were baptized. The iconic baptistery was a real architectural gem, enhanced by three Renaissance bronze doors featuring relief sculptures by Andrea Pisano and Lorenzo Ghiberti, the latter ones famously dubbed "the Gates of Paradise" by Michelangelo Buonarroti for their glittering beauty. The interior was lined with medieval golden mosaics.

Antonio was immersed into the cold holy water, and cried all the way home. That night, the Meucci family celebrated the happy event with a few friends, serving some Tuscan seasonal delicacies prepared by neighbours, along with a few flasks of excellent red wine, bought at a nearby wine store in via dell'Orto without any care for the price. Antonio's birth, however, did not make Amatis forget Napoleon's new rules based on *liberté, egalité,* and *fraternité*—liberty, equality, and fraternity. Most of them were meant to build a new, modern state in Tuscany, forcing the clergy and the aristocracy to pay taxes, along with the ordinary citizens, in order to finance public works.[2] Unfortunately, these new rules were largely ineffective because of the increased bureaucracy required to meet the standards set in Paris.

Amatis was working as a civil servant at the Palazzo Vecchio. His salary was modest, but Maria Domenica helped the family finances by knitting socks late into the night, by the dim light of an oil lamp. As a loyal government clerk, Amatis was aware of his duties. However, in the evenings after work he would often stop at his usual wine store and listen to his friends' often-critical comments.

Antonio was only a few days old when Elisa Baciocchi, Napoleon's sister, first appointed princess of Piombino and then Grand Duchess of Tuscany, arrived in Florence at the Palazzo Vecchio, even before the detailed schedule of the celebrations in her honor was finalized. But in the early months of 1814, when Antonio was just six years old, Napoleon's reign in Tuscany came to an end. Princess Elisa, along with her husband and children, left Palazzo Pitti on a cold February morning with ten carriages, stopped in Lucca for a few weeks, and then was forced to leave Tuscany altogether. As a result of the Paris Agreement in 1814 and the Congress of Vienna in 1815, Ferdinando III of the Lorraine family—now re-established as the Grand Duke of Tuscany, where he would rule until 1824—came back to Palazzo Pitti, and Florence once again became part of the Austrian Empire.

The duke's meek temperament did not offer his Italian subjects much hope of independence from the court of Vienna. The old political and social system was reinstated, with the aim of fostering a peaceful Europe after the Napoleonic Wars. To ensure the stability of Europe's pre-existing governments, Klemens von Metternich, the mastermind behind the Congress of Vienna, insisted that the old Europe should suppress liberal ideas and quench any sign of insubordination.

At the time, Antonio was attending elementary school, where he proved to be precocious and mature for his age, intelligent and imaginative, asking endless questions his teachers could not adequately answer. Among the most frequent issues, he would debate the benefits of foreign rule, and developed the conviction that his hometown should be independent. His teachers suggested to Antonio's parents that the boy's education could be accelerated, and Maria Domenica and Amatis were very pleased—also thinking that, later on, Antonio would be able to help them financially.

The whole family, in the meantime, had moved to Casa Pasqui at no. 6412, at the intersection of via de' Servi and via del Castellaccio, where they would live for many years.

When Antonio turned thirteen, Amatis decided that he should attend the Accademia delle Belle Arti near Piazza San Marco. It was a unique, prestigious public institution, within walking distance of home, and free of charge. To be admitted, applicants had to pass a very demanding entrance test. In 1821 Antonio studied intensely to meet this challenge, and on November 27 of the same year, the Meuccis received a letter with the happy news that Antonio had been accepted as the youngest student at the college.

Since the Renaissance, Florence had distinguished itself as a center of scientific research under the enlightened patronage of the Medici family. The Accademia was founded by Cosimo I de' Medici in 1563 under the influence of Giorgio Vasari, and had welcomed great artists, scholars, and scientists, such as Michelangelo Buonarroti, Benvenuto Cellini, and Galileo Galilei, as members. The campus included the Royal Museum of Physics and Natural History, the Museum of Music, as well as the Museum of Machines, where Galileo's revolutionary design for the pendulum clock was prominently displayed.[3]

The academy was meant to give students a holistic, well-rounded view of culture in its multiple artistic and scientific expressions, fostering curiosity and faith in progress through the development of useful and practical applications. No female student had been admitted to the academy until Artemisia Gentileschi in the seventeenth century, who soon distinguished herself as a painter.

During Antonio's years at the academy, he watched helplessly as four of his eight siblings died from childhood diseases. He realized how privileged he was to have been accepted by such a unique school, and observed, with great respect and admiration, how every member of the family worked so hard to supplement his father's income.

At the end of six years of attendance, Antonio would become a top scientific craftsman, equal to an engineer. His fellow students were all older than he was and belonged to distinguished Florentine families. Though this social discrepancy created quite an uncomfortable gap, it gave the boy the determination to succeed at his new school in such a respected environment.

The academy had undergone major changes in its organization during Napoleon's occupation of Tuscany (1804–1814). These involved the introduction of a number of scientific subjects along with the traditional ones, such as drawing and art history, which were taught during the first year. The second year focused on music and rhetoric, as well as the study of the French language. Antonio learned how to play the piano well and found it stimulating for his creativity. The third year, dedicated to mechanical arts, gave him an opportunity to study chemistry and mechanics, including acoustics, physics, and electricity, and above all allowed him to put his studies into practice at the nearby Conservatorio di Arti e Mestieri. From the very beginning, Antonio showed a particular inclination toward chemistry and mechanics, including the latest electrical developments, such as the emerging field of galvanism. He was constantly seeking practical applications for the theories he was learning with such great conviction and dedication.

He was fascinated by Leonardo da Vinci's scientific diagrams of flying machines with bird-like wings, which four centuries later would lead to revolutionary inventions, such as the helicopter and the airplane, then unthinkable. Da Vinci was studied at the academy not so much as one of the greatest painters of the Renaissance, known all over the world for his *Mona Lisa*, but rather as a scientist who extensively researched topics such as anatomy, aerodynamics, and hydrodynamics, as well as botany, zoology, geology, and optics.

Antonio, already a free thinker, was also intrigued by Galileo's bold astronomic discoveries that contradicted the ingrained Christian theology of his time—which was based on the belief that Earth, rather than the sun, was the center of the universe. At the University of Padua, Galileo incorporated the most powerful lenses available on the Venetian market into his newly invented telescope, which consequently could magnify by twenty times the satellites he was watching. This confirmed the pioneering work carried out by Nicolaus Copernicus almost seventy years earlier, based solely on naked-eye observations. Galileo catapulted to the forefront of the international scientific scene, despite his belief that the Vatican would be open to accepting the validity of his revolutionary discoveries, which clearly contradicted the Ptolemaic geocentric vision of the world.

Galileo's theories, for which he was tried for heresy under the Roman Inquisition and then placed under house arrest, could be demonstrated at the academy where he had been a young instructor, in particular in the observatory of its Museum of Physics and Natural History. This had become a real hub of creativity, especially under the Grand Duke of Tuscany Leopold II, son of Ferdinand III, who ruled from 1824 to 1859. It turned out to be a vital workshop for Antonio, too, affording him daily access to the Museum of Machines as well as the Laboratory for the Fabrication of Instruments of Physics.

Thanks to some of his gifted professors, Antonio was able to acquire an invaluable know-how, both theoretical and practical. Francesco Focacci was teaching mathematics and mechanics, as well as superintending the construction of machines of different kinds. Antonio Targioni Tozzetti, a renowned medical doctor and botanist, was teaching applied chemistry, by then no longer a marginal topic but a highly respected academic discipline with technical and industrial connotations. At that time, in the early years of the restoration of the Austrian rule, these protagonists of the Tuscan scientific scene were considered the best two professors in Florence in their respective fields. Moreover, Felice Gori, the masterly royal machinist, held Sunday tutorials to enrich young craftsmen in their specific disciplines.[4]

The classes Antonio attended at the academy were highly advanced, experimental, and interactive. On one hand, it was a unique pool of free-thinking intellectuals, deeply anchored in science and technology; on the other, a privileged group of students eager to learn about the most innovative subjects.

The contrasting theories on electricity from Luigi Galvani (1737–1798) and Alessandro Volta (1745–1827) paved the way more than anything else to Antonio's future research and unique discoveries. Galvani was a biologist and one of the most illustrious Bolognese scientists of the eighteenth century. Through experimentation on frogs, he studied and grasped the functions of nerves and muscles as conductors and receptors of brain stimuli. Volta, a highly esteemed professor of experimental physics at the University of Pavia, was skeptical of Galvani's animal electricity theory, proposing instead that an electrical current is generated by contact between different metals. Volta's theoretical and experimental work in this area resulted in his invention and actual construction of the first battery, proving that electricity could be produced chemically. The prevalent theory that electricity was generated exclusively by living beings was finally dismissed.

Antonio, inspired by Volta's experiments, was able to construct his own rudimentary batteries and do his first experiments by piling up zinc and silver discs, separated by paper discs soaked in acid and salt, materials available at the school. This way, a chemical reaction was triggered between the metals, resulting in a stream of electric current whose intensity could be increased or decreased. It was indeed a great achievement, even though it worked only intermittently.

Antonio was also interested in studying the effects of magnetic fields on electrical currents, the understanding of which underpinned the subsequent development of generators and motors. Electromagnetism and electrochemistry—in particular, electromagnetic induction and electrolysis—were leading fields of research and invention, with wide-ranging consequences. The sheer impact that the developments in this field had on people's lives, and on society at large, is similar to the radical innovations currently being brought about by artificial intelligence.

While Antonio was deeply steeped in his favorite studies and research, he was asked by his father to start working as well, to supplement the family's meager income. Although he considered himself a mechanic and an electrician, at that time electricity and mechanics offered few opportunities for work, so Antonio had to content himself with a less inspiring job. His sister Adelaide was twelve years old and was doing some embroidery work, while Luisa was ten and offered great help around the house. The other siblings were too young to contribute.

Thanks to the position of his father, now a senior civil servant of the Grand Duchy of Tuscany, Antonio was hired in 1824 as an entry-level guard, and later promoted to assistant gatekeeper, stationed at Porta San Niccolò, Porta San Gallo, and other gates of the city of Florence. He was essentially a customs official, stamping passports, inspecting goods, and collecting taxes. Antonio was required to wear a uniform and work eight-hour shifts, sometimes extended to sixteen hours, submitting himself to military discipline and rigid timetables. He had just turned sixteen and his salary was 40 Tuscan lire a month, equivalent to about $400 per month today.

Unlike in the eighteenth century, after the Congress of Vienna the borders were enforced and patrolled by the Grand Duchy police task force to ensure the so-called Buon Governo (Good Government) and the respect of civic order, based on strict surveillance and, if necessary, repression. The police became increasingly invasive, inspecting people and ideas more than goods. The text of documents, publications, and correspondence was thoroughly checked, and often censored if it was perceived as a threat to the government. According to the Austrian authorities governing Tuscany at that time, equal justice and fair administration could be guaranteed only by absolute monarchy. Their subjects were supposed to be subservient, unable to participate in the government or communicate with one another, except within the parameters graciously allowed by the sovereign. The G. P. Vieusseux Scientific-Literary Cabinet in particular was constantly under surveillance since it was originally established in Switzerland, providing a unique open library in Florence and facilitating an unprecedented exchange of ideas at a European level.[5]

Whenever Antonio's work allowed him some free time, he would scrupulously attend classes at the academy, where most of his professors were reluctant to accept all the new rules imposed by the Austrian authority. Starting from 1821, in fact, a deep mistrust and unrest circulated among people from all walks of life, and an informal network of secret revolutionary societies, called the Carbonari, was formed in Palermo, Florence, and Turin to fight foreign rule. Antonio became more and more involved in secret patriotic meetings, firmly convinced that Tuscany, at least, should get rid of the oppressive foreign rule.

He believed that the main goal of the Congress of Vienna in redrawing the map of Europe had indeed been reached by restoring the territorial order that had existed before the Napoleonic conquests. Nonetheless, it had failed because it did not take into account the rising nationalism across Europe, a force that would destabilize the continent throughout the nineteenth century. The Grand Duchy of Tuscany, once restored to the hands of the Lorraine family, had indeed succeeded in creating a balance of power and diplomacy. However, this seemingly peaceful state under Leopold II, who ruled for thirty-five years until the unification of Italy, was based on a stagnant status quo reinforced by the constant intervention of the police.

Antonio was arrested more than once and imprisoned for insubordination to the Grand Duke, on one occasion for hiding a stolen weapon and another for distributing flyers against the Duchy. His life became gradually more complex and demanding, especially at work. Because of his background in chemistry, he was put in charge of a team of guards assigned to the delicate task of lighting the fireworks whenever there was a major celebration. These events were planned by Leopold II's wife, the Grand Duchess Maria Anna of Saxony, who held powers similar to those of a governor.

On the occasion of the birth of one of her children, the plan was to light up for three nights in a row the Piazza della Signoria, the most important square in Florence, a symbol of the Medici power, home to the Palazzo Vecchio. The launchers, set up across the Arno river, were given eight to ten rockets each, under the direction of fireman Gaetano Baratti. He had innovatively used chlorates to obtain more intense colors, forms, and sounds instead of the usual nitrates, which were less explosive and dangerous. Baratti made a number of small pellets of pyrotechnic composition called stars, which contained metal powders, salts, or other compounds that, when ignited, make special spark effects in displays that were usually run by licensed professionals.

On the first two nights, everything went well. The whole crew was operating from a scaffolding built for the occasion, and the fireworks were very successful and widely admired by a huge crowd. On the third night, however, after Antonio lit the fuse, something went wrong. The rockets went even higher than the previous nights. Instead of falling into the water of the Arno, as planned, sparks ended up injuring those gathered on the other side of the river, causing some real damage. Eight people were injured. The police intervened immediately and arrested all three crew members, who were deemed responsible and subsequently jailed.

Antonio was the youngest of the group, yet increasingly suspected of using his position as assistant gatekeeper to express his independent spirit and protest against Austrian supremacy. He was accused of being part of the Carbonari network and of playing an important role in the struggle for the Italian unification, called the Risorgimento. Further infractions followed shortly after. Over time, Antonio had earned the reputation of being a dreamer, a careless young man rather than a diligent worker. As a result, he was sentenced to eight days in jail, three of which would be spent on bread and water alone.

While Amatis Meucci was escorting his son to prison, the police superintendent advised him to write a letter of appeal to the Buon Governo, which he did immediately. The police superintendent sympathized with both of them and made sure that Antonio received a proper meal before going into jail.

Amatis petitioned for his son's release with the following letter:

> *The Son of the Undersigned who is in Prison has not committed crimes such as to taint the honor of his Family, nor is he Guilty of any other offense if not lack of consideration and carelessness to be attributed to his youth, and which did not allow him to reflect on the damages which he has involuntarily caused; Such circumstances torment the Heart of a Father, who Humbly Implores The Goodness, The Clemency and The Mercy of your Most Illustrious*

Excellency, to ascribe to his Youth and lack of experience what the Son is Guilty of . . . And therefore Begs you to be so kind as to reduce the right Punishment to comfort an Afflicted Mother and to Forgive him the involuntary rather than deliberate shortcomings and to condescend to allow him to return to his duty as Gatekeeper, the Suspension of which increases day after day the new troubles that fall upon this Innocent and Unfortunate Family.[6]

A reply arrived immediately, accepting Amatis's proposal. Luckily, the Criminal Court granted Antonio and the others involved the benefit of the doubt. Thanks to his father's intervention and his position at Palazzo Vecchio, Antonio managed to regain his freedom in just three days by only paying a fine.

In the meantime, he managed to complete his studies in chemistry and mechanics at the academy in 1828 while helping the family finances by keeping his job as assistant gatekeeper. However, after his latest release from jail, he was regularly followed by the police because he was constantly suspected of plotting for the liberation of Tuscany. He was jailed yet again in 1829.

Antonio was a handsome young man who had conquered more than one heart by the time his boss, Gaetano del Nibbio, discovered that his lover—the very attractive wife of the owner of Trattoria del Chiù—also had a crush on Antonio. Driven by jealousy, Gaetano reported him to the Buon Governo for abandoning his post while still on duty. Accused of misconduct,

Antonio was sentenced to one month in jail and had his salary suspended. He was subsequently transferred to work at Porta Santa Croce and then to Boscolungo, a town in the Apennine mountains. The accusation was particularly painful because it was fabricated, and his father, Amatis, who had proven through the years to be a civil servant of flawless morality, had just forwarded a second petition to the Buon Governo requesting a grant to support his six children, two of whom were seriously ill, and his eighty-nine-year-old father.[7]

In spite of the positive report made by Antonio's chief in Boscolungo, in which he was praised as a serious and reliable young man, on July 13, 1830, Antonio resigned from his post without getting any support from his more deferential colleagues. His patriotic aspirations had been closely monitored. Having been rather lenient for the first fifteen years, the local police became less tolerant, and Antonio was one of the first patriots who lost his position as a result of obstinately pursuing his brave convictions. In this same period Pope Pius VII had condemned the Carbonari as a Freemason secret society and excommunicated its members.

At the age of twenty-two Antonio had to look for another job, hopefully one more attuned to his true aspirations. He turned to the theatre.

At that time in Florence there were about twenty theatres. He began to work occasionally as an assistant technical supervisor at Teatro Alfieri and Teatro Goldoni, as well as Teatro della Quarconia, established in 1650. By creating such special effects as rain, wind, lightning, and thunder, he soon realized that the theatre environment, with its magic, truly attracted him. The Quarconia was particularly popular for giving audiences the option of bringing their own food and wine to performances. The atmosphere was rather rowdy, and nights often ended with the launch of leftover bones and apple cores from the gallery to the orchestra, or from box to box. The inevitable conclusion of these boisterous and unruly gatherings was the arrival of the police and the ambulance.

In October 1833, Antonio, then twenty-five, was advised to introduce himself to the head of La Pergola, a unique historic opera house, located in the center of Florence on via della Pergola, from which it took its name. The building had been designed under the patronage of the Medici family in 1656; primarily used by the court, it was the most important venue in town, and represented a milestone in the history of theatre architecture worldwide. With La Pergola, the Italian-style theatre—with its traditional horseshoe-shaped auditorium and three rings of boxes topped with a gallery—came into existence. In 1718 it was opened to the public, and in the early nineteenth century, lavishly restored.

In this theatre, a landmark in the development of melodrama, the great operas of Mozart were heard for the first time in Italy. Composers Vincenzo Bellini, Gaetano Donizetti, and Giuseppe Verdi were given their premiere productions. La Pergola was attended almost exclusively by the nobility and wealthy foreigners staying in Florence during their Grand Tour. When the Grand Duke of Tuscany Leopold II or members of his family were in attendance, not a single noise could be heard in the opera house.

Antonio was first received by the chief stagehand of the theatre, Artemio Canovetti, whose difficult job was to set up pulleys and counterweights to move the scenery. Knowing that La Pergola was always in need of talented people, Artemio introduced Antonio to the internationally known impresario Alessandro Lanari, a leading figure in his field. Since 1822 he had successfully brought the best of Italian opera to Florence, turning his theatre into an experimental workshop, involving not only visual arts but also chemistry, physics, and optics at their best. It had become a beacon of advanced technology, recently equipped with heating. His aim was to attract different audiences, increasingly striving to provoke strong emotions, therefore requiring diversified and complex stage effects.

In spite of his short stature and grey hair, Lanari possessed a charismatic, magnetic personality, and imposed his style and working method on his employees. His life was adventurous, as he traveled constantly to hire the best singers and directors available, not to mention navigating a complicated personal life, with children from both his first and second marriages. With a touch of irony and admiration he was called "the Napoleon of the Italian impresarios" and "the greedy pirate" because he was reputed to be very stingy. He underpaid his collaborators and composers, but when he had to stage a new production, he spared no expense.

He was meticulous and fussy, controlling every detail, and this was the secret of his success. Lanari decided the theatrical program, and for each production approved the specific stage design, the sets, and the number of rehearsals before the opening night. He was equally feared and loved by his employees, and assumed the right to openly reproach them if they did not meet his high expectations. Lanari was not too tolerant of directors or critics either. He felt that even journalists had to treat him and his theatre with respect, and would not tolerate any criticism on the program or performance, since each mise en scène cost him a fortune.

He was admired as a skilled and dedicated administrator, and profitably managed the finances of La Pergola, corresponding with Giovanni Ricordi, founder of the historical music publishing company, as well as with the most important composers of the time, including Vincenzo Bellini and Saverio Mercadante, and singers of the caliber of Giuseppina Strepponi and Luigi Lablache.

When Lanari met young Meucci, he hired him on the spot, realizing he would be a great asset for La Pergola, given his previous experience. Antonio was so elated that all his hard work at the academy was finally paying off. He was being put in charge of lighting and sound at one of the leading theatrical establishments in Italy, rewarded for the first time with a dignified salary and even a private workroom behind the stage. His dream had come true. He was free to build on his hard-won education at last. No more wasting his precious time checking passports, inspecting goods, or collecting taxes! No more wearing a uniform and feeling suffocated by rigid timetables and a military discipline under a foreign authority! He had never lacked self-esteem, and that night Antonio celebrated this turning point in his life at Trattoria del Chiù with close friends, arriving home at dawn, where his mother was still waiting for him.

The following day he did not need the alarm clock or his mother to wake him up. He knew this was the beginning of the professional life he aspired to. In spite of the few hours of sleep, he was full of energy and ideas. A hopeful new future was ahead of him.

The arrival in Florence of a new generation of scientists had been instrumental in renewing Antonio's interest in keeping up with his experimental activity—in particular, how to apply what he had learned from Leopoldo Nobili's legendary twelve open masterclasses on Volta's battery. After the failed uprising of 1831, in which the distinguished patriot Ciro Menotti paid with his own life, Nobili was forced to go to Paris in exile. The same fate befell Giovanni Battista Amici, professor of astronomy, who had to leave town not to compromise his strong beliefs.

Antonio had maintained his interest in the sciences. Now, his position would give him the unique opportunity to address familiar issues such as optics, mechanics, chemistry, electricity, and physics in the context of the performing arts.

La Pergola was a modern institution, with a capacity between 900 and 1,300 people and state-of-the-art technology. Antonio was very absorbed in his work, always eager to learn everything about the technical aspects of the theatre and his new multifaceted job. He arrived on time each morning and he was the last to leave in the evening. He was disciplined, generous towards others, and quick to carry out his various duties while keeping his patriotic aspirations to himself. Antonio enjoyed solving any practical issue with intelligence and competence. For example, in order to change sets, both during rehearsals and performances, the stagehands used either large handkerchiefs or torches as signals. The audience would grow impatient during the seemingly interminable time it took to change the sets between scenes.

Not even a year after he had been hired, in 1834, Antonio invented the first acoustic telephone in history. This practical contraption consisted of two cones connected by an empty pipe. A person could speak into one end and the voice would carry through the pipe to a listener on the other. From his control room next to the stage, he would transmit orders to the stage-hands in the fly gallery and coordinate the many extras, thus eliminating the awkward and dangerous use of handkerchiefs or torches. This first device amplifying the human voice was highly innovative and proved very useful, cutting the time needed to change sets by half.

This unique mechanical invention, not yet electric, marked Meucci's first step on a lifetime's adventure, feeding the obsession with research that was to become the very focus of his existence. Overnight, it won him the respect of his colleagues and, in particular, Lanari, who considered his theatre as a think tank of innovation and Antonio his most reliable and gifted collaborator. Antonio's *tubo acustico* still exists and can be admired at La Pergola as a museum display.

During his two years at La Pergola, from 1833 to 1835,

Antonio also dramatically improved the lighting system. Candles were still used to illuminate the stage and were later replaced with gas burners between 1834 and 1850. Inspired by Galileo and the way he built his own telescope, which included hundreds of tiny mirrors, Antonio once again showed his inventive side by repositioning the existing mirrors behind each candle so they would magically disperse the light in all directions. He also worked with a newly acquired electrostatic machine, which produced additional light by generating a strong electrical charge. The audience benefited from the improved lighting effects made possible by practical technology. Though it was all invisible to the audience, it increased the power of illusion and imagination at this experimental theatre.

Antonio's duties included improving the theatre's poor acoustics, especially in the back seats. Capitalizing on his newly invented acoustic tube, he created a megaphone that amplified and directed the voices of the performers. Unfortunately, it was never used because Lanari rejected it on aesthetic grounds.

During these highly productive years at the theatre, he met Ester Mochi. From the age of fifteen she had been working full time as a dressmaker at La Pergola, assisting Lanari's sister, Isabella, who oversaw the theatrical costume atelier, located in the same building where Antonio worked. Ester helped design and tailor the costumes of each major opera singer.

Isabella, like her brother, was a perfectionist and sometimes requested that a hem or a pleat be done and redone up to ten times. This required meticulous handwork, since sewing machines had just been invented and were not widely used. Unlike Lanari, Isabella was amiable, always smiling, and treated her assistants with great respect.

Ester, with intense black eyes and long raven hair neatly braided, was twenty-three when she met Antonio, who was two years older. Whenever he would walk by the atelier, their eyes would meet. He later noticed she had moved her chair next to the open door of the main exit, so she could have a better view of him passing by.

One day Isabella asked her to show some fabric to her brother, the great impresario, for his opinion. While rushing to his office, Ester came across Antonio on the staircase, who asked her why she was in such a hurry and expressed the wish to know her better.

After that fortuitous encounter, Antonio thought of her all the time. He could not eat or sleep, to the astonishment of his parents and siblings. While at work, both of them sought every occasion to bump into each other. In the evening Antonio would wait for her after work, not in front of the building but discreetly a few yards away, because she did not want to be seen by her colleagues. They would walk together for a while and exchange ideas along the Arno river up to via dei Serragli. However, she would not allow him to walk her all the way home, where inevitably he would have met her parents.

He was rather introverted, so it was not easy for him to express his feelings, even though he often told her he wondered why he had not met her before. They both came from poor families in the San Frediano district, and both started working at a very young age to supplement the household income before being hired by La Pergola, thanks to their professional expertise.

Ester's mother, Assunta Papini, gradually allowed them to meet and go for walks, during which they would share their dreams with each other. One evening on the Ponte Vecchio, Ester unexpectedly told Antonio they should get married. Since they loved each other, she figured, why waste time? She had already informed her mother of her intentions. He replied that he had not yet been able to find the courage to ask her, but he agreed wholeheartedly.

Back at home, Antonio's mother, Maria Domenica, noticed how silent and particularly pensive her son was. When she asked him about it, he told her he intended to get married. So, as custom required, his mother sent her husband, Amatis, to pay an official visit to Gaetano Mochi, the father of the future bride. Gaetano was wondering whether Antonio was that same young man who often ended up in jail for patriotic reasons, but admitted that his daughter was in love and eager to fulfil her dream and marry him.

The wedding preparations took all autumn and winter. Every Sunday Antonio and Ester went for long walks in the Boboli Gardens, and often he would extravagantly hire a carriage to go out of town, to Poggio Imperiale. As Ester was saving every cent for their new life together, she was concerned by Antonio's generosity. She would reproach him and sometimes even cry that she feared living in poverty again, as she did in her childhood. They managed to find a small, lovely furnished apartment at a reasonable price in the Fondaccio di Santo Spirito area. Finally, the date of the wedding was set for August 7, 1834, in the grand Santa Maria Novella basilica in Florence.

After the ceremony, Ester insisted on prolonging their ride in a rented carriage up to the nearby town of Fiesole. The two best men paid for the unexpected expense. A lovely reception and an intimate wedding banquet followed until the early afternoon, and then the couple retired to their new apartment for the honeymoon for two days, which was the maximum leave that Lanari allowed.

Back at work, Antonio was also helping—unasked—in other departments as a problem solver, even in the atelier where his wife was working on costume design. He'd often be asked to accompany Lanari out of town, and would also spend hours in the control room or in the fly gallery, where he was constantly refining his new invention and testing it. He sometimes even involved the singers, disturbing the orchestra rehearsals and upsetting his direct boss. His colleagues gradually lost interest in this acoustic research and thought Antonio was just obsessed with this contraption while he gradually became uneasy at work.

Even at home he involved his wife in testing, trying for hours to see if the acoustic tube would work both ways. One of them would speak into it while the other would listen, and vice versa. He was convinced that his invention represented a major breakthrough and that one day they would become rich and famous. Indeed, speaking tubes like Meucci's opera house invention would soon start to be used on ships and in houses, allowing people on different floors to communicate with one another.[8]

Meanwhile, Antonio was experiencing further discomfort at

work because an anonymous letter denouncing him—suggesting he had never given up his patriotic aspirations—was sent to Lanari's attention.

In fact, before being hired at La Pergola, Antonio was involved in the conspiracy for the liberation of Tuscany in 1833, and had served months in jail with the well-known patriot Francesco Domenico Guerrazzi for joining Giovane Italia (Young Italy), founded by Giuseppe Mazzini. The aim of the new movement was to promote a general insurrection in the Italian reactionary states. Mazzini was convinced that a popular uprising would ultimately create a unified Italy. The movement garnered about 60,000 convinced patriots, including humble workers, young lawyers, scientists, artists, and many members of the Carbonari, previously dispersed and ostracized by the Church. The most famous member of Young Italy was Giuseppe Garibaldi, who, once sentenced to death, chose to go into exile and keep promoting Mazzini's enlightened vision.

From the moment Lanari received the letter, the surveillance on Antonio was relentless, and the Austrian authorities, ever more viciously repressive, labeled him as persona non grata to the Grand Duke. Antonio was explicitly advised, through his father, to leave Florence to avoid being jailed again. That same year the Vieusseux Cabinet was heavily censored until it was forced to close down, marking the end of an era of open-mindedness, free thinking, and fruitful circulation of ideas.

Antonio received a lucky break. In the fall of 1835, thanks to a highly appreciated introduction from Lanari, Antonio was given an opportunity to help establish Italian opera in Havana.

Renowned Catalan impresario Don Francisco Marty y Torrens offered him and his wife a five-year contract. The Meuccis were glad to start a new chapter, however challenging it might be.

Before Antonio left Florence, he reaffirmed his affiliation with the Freemasons as well as with Giovane Italia—which had just expanded into Giovane Europa, an internationally oriented association of committed patriots that, in the years to come, would lead to the unification of Italy. With their support, Antonio hoped for more appealing living conditions on a far-away island across the ocean such as Cuba.

He felt truly privileged to be able to escape the haunting surveillance of the local Florentine police by accepting, at just the right moment, an appealing contract as chief engineer—along with his wife as chief costumer—at Havana's Gran Teatro de Tacón. The theatre, soon to be built and dedicated to the governor, Don Miguel Tacón y Rosique, was going to be the largest in the Americas, as well as a prominent cultural center in the New World.

Chapter Two

INVENTING IN HAVANA
(1835–1850)

When Antonio Meucci was pressured to leave Florence, he was twenty-seven, a fervent patriot ready to give his life for a future united Italy, a convinced follower of Giuseppe Mazzini's ideals, and a particularly accomplished theatre chief engineer. His solid technical and scientific background had shaped him as a man and as a professional inventor. He had a lot to offer in his demanding new assignment in Cuba, then still under Spanish dominion and ruled by a governor. Moreover, the presence of his wife next to him, with a well-paid independent contract as costume designer and dressmaker, represented the perfect, unexpected occasion for both of them to advance their careers. If he had stayed in Tuscany, he would have been persecuted for his political ideas, most likely ending up in jail again or being compelled to emigrate, perhaps to Paris (since he was fluent in French), like other scientists he knew and admired.

The departure from Florence was scheduled for October 6, 1835, by carriage. He and Ester crossed the countryside on dirt roads, changing horses every few hours, reaching the coast of Tuscany by the following day, and then the port of Livorno on the Ligurian Sea, where a three-masted vessel under the flag of the Sardinian navy was waiting for them. Antonio had seen the sea only once or twice, and always from far away, when he was asked to accompany Lanari on business trips either to Naples or Senigallia. Ester had no idea what to expect from her first encounter with the sea. Deep down she feared it would change her life forever.

The *Coccodrillo*—once a pirate ship, then mainly used to transport coffee, sugar, and rice from Cuba to Europe—was usually empty on its way to Havana. This time, it hosted quite a number of passengers, including singers, orchestra musicians, tailors, technicians, and other employees hired by Don Francisco Marty y Torrens, some of whom—like the Meuccis and other patriots—were traveling without a passport.

Antonio and Ester had bid farewell to their families the night before. Their intention was to remain in Cuba for the duration of their five-year contracts and then come back to their beloved Florence. They were pleased that renowned tenor Lorenzo Salvi, who would become a lifelong friend of theirs, had joined them in this adventure of promoting Italian opera in the new continent. Crossing the Atlantic without a steam engine would take quite a long time, and during that risky major voyage together, they became like family. They were all looking for a better life, sailing into a promising future.

The ship's cargo included some thirty-five tons of props and equipment to set up the opera stage in Havana, as well as eighty-one special passengers whose presence was deliberately undeclared to the Austrian authorities. This was a perfect setup for the Meuccis to avoid the persistent surveillance of the police of the Grand Duchy while waiting for better times. The Giornale di Commercio of Porto Franco, Livorno, reported the ship's departure on October 7 without mentioning any passengers.

The journey across the Atlantic took two and a half months, since the ship relied only on its sails and followed the route of explorer Christopher Columbus. After the first stop in the Bay of Gibraltar, located at the southern tip of the Iberian Peninsula, the brig crossed into the Atlantic Ocean via the narrow strait, which bordered Spain to the north and Morocco to the south. After ten days or so, the ship moored near the coast of northwest Africa at the Spanish-owned Canary Islands. For the second leg of the trip, the captain wisely chose to continue on the same route that Columbus took. He felt somewhat protected from possible plundering pirates thanks to the cannons installed on each side of the cargo ship. That was the longest stretch of the journey, with no stopovers foreseen.

The ship was vulnerable to storms, and its original sails, though frequently mended, had never been replaced. The *Coccodrillo* was not a passenger ship and therefore accommodations were basic, to say the least. The cargo ship was equipped with large, rudimentary cabins with bunk beds. Men had separate dormitories from women with children. Climbing steep gangways and internal staircases required special attention.

Each day would start at 7 a.m. with a bell indicating that barley coffee was being served. Two simple meals a day were provided. The captain was a man of few words. He occasionally walked on the deck, hands clasped behind his back, giving orders to the sailors and constantly monitoring the variable conditions of the sea. In those days, most passengers going from Tuscany to Cuba were hoping to find work as day laborers at the coffee and sugar plantations. But Antonio and Ester were among the privileged passengers who were going to play a major role in the founding of a new grand theatre in Havana and, above all, already had work contracts in their pockets.

The journey across the Atlantic was rather trying because of the unpredictable weather conditions. It was a test of endurance. Ester felt seasick most of the time and lost weight. Her deep sense of nostalgia for Florence was compensated by the hope of a more rewarding life on an island in the Caribbean Sea. In spite of her poor health, she kept busy sewing while Antonio wrote in his diary, intrigued by the atmospheric humidity at sea. After consulting a few books, he started experimenting with substances available on board, and many years later, in 1876, in New York, he patented his own hygrometer—an instrument used to measure humidity.

At sunset on December 16, 1835, the brig reached the shores of Cuba, entering the Canal del Puerto in Havana. From far away the island appeared in all of its green splendor. The Meuccis experienced a deep sense of relief and gratitude. On December 17, the local newspaper *El Noticioso y Lucero*, announced with excitement the arrival of the ship, proudly mentioning its eighty-one passengers from the Compagnia Italiana (Italian Company). A new chapter for Antonio and Ester, as adult professionals, was about to begin under the protection of Don Francisco Marty y Torrens.

The ship's passengers looked out at their new city. Beyond the ancient city walls and bastions, the majestic colonial churches contrasted with the typical low stone houses; the city streets were parallel to one another and from a distance they suggested an orderly chessboard. Everybody on board felt a sense of curiosity and wonder. They immediately realized that the warm temperature was so different from the chilly winters they were accustomed to in Florence. Their clothes were far too heavy for the new tropical climate. The air was pleasantly filled with the scents of lush vegetation and amazingly colorful flora.

When Christopher Columbus first arrived in Cuba on October 27, 1492, he was reported to have exclaimed, "*La más hermosa tierra que ojos humanos hayan vieron*" ("the most beautiful land humans ever laid eyes on"). He was convinced that he was en route to Asia; instead, he landed in the Caribbean islands by accident. He discovered an island inhabited by over 50,000 indigenous people who, through disease and the Spanish imperial rule, would soon be largely eradicated or absorbed by intermarriage.

Throughout the centuries after Columbus's arrival, the Caribbean became an area of contention for competing empires, particularly England, France, and Spain. When Napoleon occupied Spain in 1808 and crowned his own brother Joseph Bonaparte king, all the Spanish colonies grew hostile toward the French. In 1810 they all revolted, except Peru and Cuba, where the Spanish governor at the time, Salvador José de Muro, managed to keep his government in power. When the French occupation of Spain ended in 1814 and Fernando VII was restored to power—and even more so ten years later, when the only colonies left to Spain were Cuba and Puerto Rico—it became customary to call Cuba "the forever faithful island." By then, the whole island had developed a dependence on its sugar plantations as well as on slaves, imported from Africa since 1511. Their presence gradually grew to outnumber Cubans of indigenous or European origin. In a world steadily moving toward the abolition of the slave trade, Cuba was reluctant to follow suit.

Don Miguel Tacón y Rosique had been governor for a year when the Meuccis arrived. In almost no time he had radically transformed Havana and the entire island, enabling the colony to achieve great progress in many fields—with the exception of political freedom. Among his most important innovations was Havana's state-of-the-art public lighting system, which caused a sensation among inhabitants and tourists alike. Over 500 glass streetlamps had been installed all over the city. As Antonio disembarked the ship, he immediately noticed their innovative presence and the rare beauty of the pinkish glass refracting the light at sunset and dawn. Quite an unexpected sight for a far-off island in the Caribbean.

Cuba turned out to be a safe island. In fact, a very efficient night patrol, reaching their post one hour before sunset and leaving after dawn, had been recently established. The cleanliness of the streets of Havana, still unpaved, was maintained by having them properly swept every day.

Many horse-driven carriages and mule-driven carts were waiting at the port to pick up members of the Italian opera company and to load their baggage. Antonio and Ester noticed that only carriages were going by; no people were walking. Later, they realized that only slaves would walk in Havana because nobody, especially ladies, would ever leave their *volantas*, or flying carriages (so called because of their proverbial speed), not even to shop or stop for a drink or ice cream. Black slaves would attend to them as they sat comfortably, sparing them the unpaved roads, too muddy in winter and too dusty in summer.

Governor Tacón had set very strict rules for passengers disembarking from ships that arrived in the harbor, but since Don Francisco Marty y Torrens had personally vouched for the whole Opera Italiana company on the *Coccodrillo*, the procedure was drastically simplified. In addition to being the owner of the new theatre, Don Francisco owned an immense trading empire on the island and was highly respected, if not revered. So the government officials boarded the ship and simply ticked off the names on the list they were given, without even asking for any documents.

Upon arrival in Havana, all the artists received an invitation for the following night to a formal dinner hosted by their sponsor, Don Francisco, at the Fonda de Neptuno. It was an expensive hotel in Plaza del Vapor, a central square originally used as an open market. On the same strategic site, as part of the architectural complex, stood the Pescaderia, a neoclassical-style, two-story gem of masonry with marble floors and fine woodwork. It was recently built, entirely at Don Francisco's expense, in exchange for a twenty-year monopoly on fishing. This extraordinary deal was made possible thanks to his excellent relationship with Don Miguel Tacón, who recognized his entrepreneurial merits.

Born in Barcelona, Don Francisco had emigrated to Cuba at the age of twenty-four and became wealthy through his involvement in the slave trade, among other ventures, which was then considered a patriotic activity because it supplied cheap labor for public works. He belonged to the particularly creative, industrious Catalan community who controlled wholesale and retail trade in Havana, and made a fortune in a relatively short time.[9] Moreover, he had the great foresight to bring, at his own risk, an exceptional cast of singers and set designers from Florence and conceive the Gran Teatro de Tacón as one of the largest opera theatres in the world at the time, instilling this way the appreciation of Italian *bel canto* in the Americas.

Don Francisco had invited a number of wealthy patrons to the dinner in honor of his guests, and arranged for Don Manuel Pastor, chief mechanical engineer and inspector of the island's fortification, to sit next to Antonio Meucci. They could comfortably converse in French, which at the time was the international language of culture. This gave Don Francisco the opportunity both to introduce Meucci to some influential people in town and help Don Pastor discuss a few urgent technical challenges of the city. Since Meucci came from Florence, the very cradle of the arts and sciences, and was educated at the prestigious Accademia delle Belle Arti, perhaps he could help solve them.

Pastor was an excellent civil engineer but, unlike Meucci, he had no knowledge of chemistry. He asked Antonio for advice on how to fix the poor quality of the water in Havana. Later, after doing a thorough chemical analysis, Antonio discovered the water was hard, containing a high concentration of calcium and magnesium ions. So he convinced the municipality to change its entire filtration system. As a result, the water in Havana became soft and pleasant to drink after the new filters had chemically neutralized the corrosive effects of calcium and magnesium. Moreover, the water pipes were no longer coated with crusty deposits and the market demand for spring water decreased.

After this successful intervention, Antonio's reputation was greatly enhanced, and he became well known in town as a problem solver at many levels. Lots of prominent people came to him for advice.

Antonio's top priority was the construction of the Gran Teatro de Tacón and the adjacent building, including the costume atelier and workshop, as well as a laboratory for himself and his wife. The building permit was granted by royal decree in Madrid two months after they arrived, and the construction started in July 1836, entirely at Don Francisco's expense. He would then become the owner of the theatre, receiving in return a twenty-year monopoly from the government, as had successfully happened before with the Pescaderia. In other words, no other theatre in town would be able to compete with him, but there was still a lot of work to be done.

The Gran Teatro de Tacón would rise from a plot of land by the botanical gardens, just outside the city walls and near the modern, highly functional railway station, which was inaugurated in November 1837 and allowed the new trains to transport sugar, coffee, and rice from the production centers directly to the port of Havana. Antonio was very impressed that the main railway line in Havana had preceded the first railway in Spain by more than ten years.

Don Francisco counted on Antonio's rare expertise in advanced strategies in theatrical lighting and, above all, in acoustics. From the very beginning, Antonio experienced a sense of trust and a great freedom, allowing him to express his creativity in fields he had never addressed before.

First of all, Don Francisco asked him to review the drawings and blueprints, assuming that Antonio and Ester would reside in the adjacent living quarters. Antonio became immediately concerned with the architectural structure of the building, far too large for an opera house. Despite its huge size, its basic layout was very similar to that of La Pergola in Florence, with the royal box located directly facing the stage. On the ground floor, however, there would be only iron gratings, and on the upper floors, eighty windows with no panes, only shutters, that had to be kept open for ventilation. The voices of the singers would get totally lost in such a space, as large as an arena.

Luckily, Don Francisco accepted Antonio's daring proposal to channel part of a nearby subterranean river into the basement of the theatre, thus creating a resonance chamber and reducing the volume of the theatre. Antonio's observations and proposals were extremely astute: Devising a way to incorporate a flow of water into the resonance chamber would drastically improve the acoustics. His work preceded the discoveries of American inventor Joseph Henry in this specific field by fifteen years.[10]

The Gran Teatro de Tacón officially opened on April 15, 1838, and as Governor Don Miguel Tacón entered the theatre, which was dedicated to him, the royal march was solemnly played. Once completed, it was the third-best opera house in the world, after La Scala in Milan and the Vienna Opera. Antonio and Ester moved into the annex, where they had a large, comfortable apartment. As promised, there was a laboratory/workshop for Antonio and a costume atelier for Ester on the ground floor in the left wing of the main building—a reward they never would have dreamt of in Florence.

Throughout the two years preceding the opening of the Teatro Tacón, Antonio was totally absorbed by the new theatre requirements and at the same time felt truly blessed, since he had the opportunity and the means to let his creativity run free in many directions. The artists had nothing to worry about either: Don Francisco was also the owner of two more playhouses—the Diorama for dramatic performances, and the Principal, which was mainly dedicated to the opera but had become far too small. In January 1836, the Nuova Opera Italiana had its debut there, highly acclaimed by the local press, staging *The Capulets and the Montagues* to commemorate Vincenzo Bellini's premature death.

To avoid yellow fever during the hot, rainy season in Havana, the company was, from its very arrival, profitably scheduled to perform in Mexico City, Santa Cruz, New Orleans, and other locations. Upon request, it would also perform elsewhere in the country—for instance, in Santiago de Cuba during the spectacular and highly attended Carnival festivities. But the Gran Teatro de Tacón gradually became the opera company's only venue, especially after 1846, when the Principal and the Diorama were both completely destroyed by a sizable hurricane and never rebuilt. The Tacón was also damaged, but was quickly—if temporarily—repaired, just in time to present Verdi's *Ernani*, featuring six new singers along with musician Domenico Mariani, all hired from Milan by Don Francisco.

In 1847 the venue was accurately renovated and updated under Antonio's charge—by this time he was entrusted with the direction of the works. Under his leadership an innovative mechanism was introduced to raise or lower the stage quickly, a new ventilation system was set up, and toilets were added, a feature that did not exist in any other theatre in town. Antonio installed his well-tested acoustic tube and used it on a daily basis, as he had in La Pergola. He did not have a boss to report to. Unlike in Florence, he was the person in charge of the entire management and the production of every performance. His professional life was marked by a pronounced spirit of service, which kept his curiosity and constant compulsion to experiment alive.

Life and work there had a strong Italian accent. The prima donnas were all Italian, as were their male counterparts. The rest of the theatre personnel also came mostly from Italy. Such groups of opera singers often formed an island unto themselves. They remained faithful to their own traditions, usually without mixing with the local people. Since they spoke mainly among themselves, they often never learned the language, even after years in a foreign country.[11] But Antonio and Ester had both become fluent Spanish speakers.

Though they were very busy at work during the day, the Meuccis loved to have people over in the evening, and their home gradually became a magnet for socializing, a very special meeting ground. Friends would stop by for a drink or to smoke a cigar and enjoy stimulating conversations.

Despite the torrid climate, most Spaniards fully enjoyed life and had very few hours of sleep. They had long working days, often starting as early as 5 a.m., with lunch at 3 p.m., then a one-hour siesta, followed by more work and an intense social life at night. The Meuccis felt at home on a happy island surrounded by friendly and outgoing people who truly respected them.

Antonio, however, could not help thinking of his own parents back in Florence, always working hard to feed the family, while he and his wife lived in a sort of paradise, where a great variety of exotic fruit and vegetables was available in abundance, as were the best coffee and tobacco, not to mention rum, the national beverage of Cuba.

Ester, too, was constantly thinking of her parents while she was designing different sets of costumes for the opera singers of the Tacón Theatre, and sewing them one by one by hand (sewing machines would only be introduced on the island some ten years later). Her job as a dressmaker allowed her to behold and enjoy beauty and harmony by choosing the most refined fabrics, including the best Italian silk and Indian cotton. She gradually became close to the prima donnas she was dressing, who, in appreciation for her dedication and work, gave her exquisite jewelry that she would treasure.

Moreover, since the theatrical season in Havana was from October to the end of April or May, Antonio was able to spend a considerable amount of time keeping up with the latest discoveries in chemistry and electricity. Inspired by the publications of Italian chemist Luigi Brugnatelli, the first to invent the process of electroplating, he became more and more interested in this new technique, using nonprecious metals such as iron, copper, or bronze with gold or silver.

In 1844 Ester and Antonio were blessed with the birth of a daughter, whom they called Juanita, who made their lives together even more joyful and meaningful.[12]

That same year Antonio received a unique proposal from General Leopoldo O'Donnell, who had become governor of Havana in September 1843, succeeding Jerónimo Valdés, a very mild and peaceful leader. O'Donnell had been chosen by the government in Madrid for his despotic nature and iron fist. This soon led him to order a bloody repression in Havana, mostly of Creoles and people of color, which caused a major exodus to the United States.

Though he was an authoritarian Spanish military general, O'Donnell loved the opera and considered Antonio a wizard, capable of creatively solving any problem. So, instead of spending a fortune to send all his soldiers' metal equipment to Paris or to London for electroplating to prevent rust, the governor asked Antonio if he could carry out the whole process locally. Antonio willingly accepted the challenge.

Using the large laboratory he had at his disposal, he employed a dozen workers to electroplate the swords, helmets, and other military supplies as part of a four-year contract directly with O'Donnell. All he needed were galvanized batteries, which he ordered from Gaetano Negretti, a friendly and reliable Italian importer based in New York. Negretti's brother would ship them, and Gaetano would deliver them in person to Havana, sustaining his ongoing business relationship with the Tacón Theatre. Cases would regularly arrive filled with Bunsen batteries, insulated copper wire, and other equipment that thrilled Antonio.

Though it's unknown today which metals Antonio used for electroplating, nickel would have been the most effective and inexpensive anti-corrosive coating for the Cuban helmets, swords, buckles, and even buttons. Not only was Antonio performing this service for the Cuban government, but even his neighbors would line up outside his laboratory to have their household goods, silverware, and candelabra electroplated for years to come. It became an inventive and financially gratifying after-hours occupation for Antonio.

Almost without realizing it, Meucci had introduced electroplating to the New World as a result of his constant experiments. His workshop in Havana was, in fact, the first of its kind in North America.

Antonio's popularity after this latest success had grown to such an extent that a special evening was organized the same year, on December 16, at the Gran Teatro Tacón to honor his many talents and eclectic personality. He did all of this while he was still in his thirties!

The Meuccis were earning considerable sums. Their five-year renewable contract with Don Francisco came with a salary of about $3,000 a year, paid in pesos fuertes, in addition to free lodging and a minimum of two slaves as house help. Considering the high cost of living, such a salary was not exceptional; however, it was almost eight times more than what Antonio had received from the Buon Governo in Florence. He also earned about $60,000 (paid in *pesos fuertes*) from an impressively large four-year contract with the army for electroplating; the equivalent today would be well over $1 million (though he also had to pay many work-related expenses out of this sum). In addition, both Antonio and Ester received gifts from actors, actresses, and singers for various personal services.[13]

Thanks to his close friend Lorenzo Salvi and the importer Gaetano Negretti, Antonio was able to keep up to date with recent publications such as the *Traité expérimental de l'électricité et du magnétisme*, in seven volumes, by A. C. Becquerel, published in 1840 by the Librairie de Firmin Didot Frères of Paris. His great interest in Luigi Galvani, considered the father of electrophysiology, led him to rediscover Galvani's *Commentary on the Forces of Electricity in Muscular Motion*.

He was also particularly interested in the theories of animal magnetism by Franz Anton Mesmer, whose *Traité de magnetism animal*, was published in Paris in 1778. Antonio enjoyed reading and rereading it in French, since he spoke the language fluently. He was particularly aware of the success that Mesmer's therapeutic treatment, based on natural and artificial magnets, was gaining all over Europe. He avidly studied the workings of nature, but his approach to natural phenomena was, above all, empirical. So when Antonio read that illnesses could possibly be treated with electricity, he felt he simply had to try it.

Based on these various theories he was able to build a rudimentary device for the cure of arthritis, rheumatism, and migraines. His house became a de facto laboratory, known to the locals as the "House of Miracles" or "The Health Place," recognized by the government of Cuba under Governor O'Donnell, which made it accessible to everyone. Meucci developed a way to treat illnesses with electric shocks, which was already a very popular medical treatment. As a matter of fact, electrotherapy throughout the nineteenth century was applied with beneficial analgesic effect. Though Antonio had no medical credentials, he was known to be working at the forefront of many different uses of electricity and was consulted by some Cuban doctors.

While experimenting to discover what beneficial effects electricity could have on physical pain, he unwittingly created an aura around himself—just like anyone working with electricity in the mid-nineteenth century did—as a person endowed with uncommon powers, a near-magician. His lab was his realm, outfitted with the most technically advanced equipment of the time, which he ordered from England and the United States.

He himself described in detail his first experiment in electrotherapy in Havana in 1848–49, illustrating it with a clear drawing that would end up being used years later in his court testimony in the Bell-Globe trial in New York.[14] A former employee came to his house-laboratory, tortured by a terrible headache. Antonio placed two cork-insulated copper spatulas, communicating via a copper wire with Bunsen batteries, in the patient's hands. He was three rooms away in his lab with his batteries when he asked his patient to put one spatula in his mouth.

What happened next was a crucial moment in the history of science. While putting the copper part of the spatula between his lips, the patient emitted a loud cry which Antonio, while holding an identical instrument to his ear, heard. He sensed right away he was onto something major. He had stumbled upon the possibility of using electricity to send the human voice through a wire.

Now he could take his speaking tube invention to a whole new level. Antonio realized for the first time that sound, and thus the human voice, could be carried by electricity through wire. Although it can't be said that in 1849 Meucci invented the phone the way we envision it today, he definitely grasped its basic principle, and for years to come, he would work relentlessly on its improvement.

At that time, instant communication over a distance was possible only by transmitting written messages by telegraph, the groundbreaking invention by Samuel F. B. Morse only five years earlier, in 1844. The first message, using the alphabetical code of dots and dashes, was sent from the Capitol in Washington, D.C., to Baltimore, Maryland, and the credit for the invention, though fiercely contested by scheming rivals, finally went to Morse, thanks to his obstinate tenacity. In those years the telegraph seemed like such a miracle that it was almost impossible to imagine a better device. The transmission of the human voice seemed pure fantasy at that time, but from that moment on, Antonio set himself the task of going a step further and constructing a "speaking telegraph."

He began a series of trials aimed at improving upon his first result, which was conducted using a 114-volt battery. To prevent electric shock, he covered the patient's instrument as well as his own with a funnel made of pasteboard. He repeated the experiment many times, first detecting an inarticulate sound. He decided to reduce the electrical charge and got increasingly better results.

Meucci therefore became the first pioneer of the telephone, or speaking telegraph—the *telettrofono*, as he first called it—as an accidental consequence of his experiments in the field of electricity applied to medicine.

Many more patients, pleased with the results of the electric shock treatment, came to his lab, the only one on the island. Their frequent visits turned these first attempts of electrotherapy into a lucrative and exciting side activity.

From that moment onward he continued his experiments, replicating and perfecting his new method of voice communication. He gradually exchanged his rudimentary transmitter for an electrostatic transmitter in the mouth as well as an electrostatic receiver placed at the ear. In this way the exact word could be clearly heard.[15]

Though Antonio would later produce a proper "static

telephone" in 1865 in the U.S., the confirmation that his first prototype actually worked took place on New Year's Eve 1850 at the Meuccis' residence in Cuba.

On that occasion, their best friend, Lorenzo Salvi, sang a few improvised Italian arias from their master bedroom while Antonio placed wires to the living room, where guests were invited to listen to his performance. The event attracted a number of cheering people, who, from that moment on, considered the Florentine inventor the pride of Havana for having conceived and constructed this remarkable device—one that would later become an unprecedented instrument of communication for the entire world. At this time Alexander Graham Bell, who would later be given credit for the same invention, was just two years old.

Until the tragic loss of their young daughter due to illness, the Meuccis spent the brightest fifteen years of their lives in Havana. Both of them prospered thanks to a solid and rewarding contract, twice renewed, allowing them to enjoy life while working hard. They were hosted in a comfortable residence far away from the hardship Antonio endured in Florence, were blessed with a pleasantly mild climate that was perfect for Ester's delicate health, and received constant encouragement and support from special friends.

When the twenty-year government concession for the Tacón Theatre was about to expire, Don Francisco gradually withdrew from his enterprise in Havana. Cubans had lost their initial interest in the opera, so running such a grand theatre was no longer a profitable business. Don Francisco started diverting his interest in organizing opera performances from Cuba to the U.S., in particular at the Castle Garden in Battery Park, Manhattan, where the genre was highly appreciated. He planned to build on the influential legacy of Mozart's Venetian librettist Lorenzo da Ponte, who had moved to New York City in 1805 and founded the Italian opera house there—which, although it had to close after two years, proved to be a true predecessor of the New York Academy of Music and the New York Metropolitan Opera.

At the same time, Antonio was looking for a more vibrant and dynamic place to develop his experiments, so leaving Havana became a stronger option every day. Many Italians began to settle in California, attracted by the Gold Rush of 1848, and many others emigrated to the East Coast, scattering all over the New York region and settling in Brooklyn, the Bronx, the nearby towns in New Jersey, and mostly in Manhattan. According to the Meuccis' best friend, Salvi, the only way not to end one's life on a happy island without a future was to land a few miles away, on the American coast.

Ester did not agree. Her first choice would have been to return to Florence, to finally be among people she knew and loved, especially now that their only daughter, Juanita, had died suddenly of yellow fever at the age of six and had to be buried in a mass grave. Antonio remembered with grief his parents back home, who had lost three young daughters, Maddalena, Giuseppa, and Assunta, and one son, Roberto, all before the age of five.

But the political conditions in Florence were not yet favorable for the Meuccis to return. In 1849, thanks to the Austrian army, Leopold II had come back to Tuscany from a voluntary exile of a few months in the Italian town of Gaeta, and drastically revoked all previous concessions, including the freedom of the press. He even restored the death penalty for acts of violence against the government. Moreover, after the death of King Carlo Alberto of Savoy, his successor, Vittorio Emanuele II, was forced to sign a treaty with the Austrians in Milan. All aspirations to Italian independence were crushed, and Giuseppe Mazzini's vision of a united Italy became a mere dream once again.

Some members of the Gran Tacón company decided to stay in Cuba, while others, like the Meuccis' friend, tenor Lorenzo Salvi, chose to go New York; he would perform there first, and then go back to sing again at La Scala in Milan.

The Meuccis' sponsor and friend, Governor O'Donnell, encouraged the couple to leave Cuba, making a convincing argument that while Havana was the present, America represented the future. Moreover, life on the island became more problematic and vexing when, in 1848, O'Donnell was replaced by despotic governor Federico Roncali, the Count of Alcoy. People were now sentenced to life imprisonment for treason against the Crown of Spain if deemed guilty of any attempt of insurrection. In September 1849, a Junta Cubana was established in the United States with the aim of instigating a revolution to liberate the island, and General Giuseppe Garibaldi was regarded as the great hero inspiring the Cuban independence movement.

Don Francisco, though Spanish due to his upbringing in Barcelona, decided the best thing to do was to leave Cuba. He chose to depart quietly, organizing a number of tours for the whole crew, starting from New York in close partnership with his well-respected Austrian colleague, impresario and conductor Max Maretzek, who was then the director of two major theatres in Manhattan, Castle Garden and Astor Place.

Given the circumstances, Don Francisco would not be renewing the Meuccis' contracts at Gran Tacón for the fourth time—nor that of any other company member—because it was no longer economically feasible. Above all, he also strongly felt it was time for Antonio to leave Cuba, especially on account of his republican ideals. In 1849 Meucci had aided Garibaldi and other prominent Italian patriots with large sums of money for the siege of Rome, and his donation was well known to Governor Roncali. Antonio realized that after fifteen years of peaceful life, his deeply felt and generous patriotic spirit was working against him once again and could easily put him and his wife in real danger.

It was another major turning point in their lives. Since Florence was still under the ever more stringent Austrian rule, Antonio, now forty-three, had only two options: stay in Cuba in spite of its increasingly corrupt and unstable government, or leave the island for the United States. In Havana, the Meuccis were a highly respected part of the community. In the U.S., however, they would have to start all over, without knowing any English and without any prospect of salaried employment. It was an excruciatingly difficult decision, since Antonio's creativity had been highly recognized in Cuba, especially after his recent invention of the speaking telegraph: his telettrofono.

Eventually, Antonio decided that he and Ester would leave the country—and take their chances in the United States.

Chapter Three

STATEN ISLAND, NEW YORK (1850–1871)

Meucci's bold decision to move to the U.S. was mainly driven by his strong conviction that New York City was the place where he could finally commercialize his telettrofono invention and make a living from it. He already knew he would find better material for his experiments there, such as batteries and laboratory equipment. He also counted on his family's savings, which were higher than he had initially realized: a nice nest egg of about 26,000 pesos fuertes, or $30,000—equivalent to about $600,000 today—thanks to Ester's careful financial management and a substantial number of valuable jewels she had received as gifts in Havana.

Ester, despite the fact she was illiterate and signed her documents with a cross, was well aware that she herself had put aside almost $13,000 in fifteen years of hard work. She was determined to protect her savings from Antonio's extravaganzas supporting a free, united Italy.[16] She knew that in New York City there was a strong Italian community of fervent expatriates, exemplary in their sense of brotherhood and solidarity, with grandiose patriotic plans sure to tempt her husband. She also knew that Antonio often got carried away in costly experiments that did not necessarily lead to profitable results. So, before they left Cuba, she was wise enough to have Antonio sign a written declaration in front of a witness, stating that at least her share of over $7,000 (plus a bill of exchange of $3,500 for a loan to their friend Lorenzo Salvi) would be deposited into a local bank upon arrival in New York. This would be sufficient to immediately buy a property of her choice, thus fulfilling her dream.

Ester understood that, all things considered, it was logical for them to follow most of the Havana opera company to New York, where, in the last thirty years, opera had become the most prominent form of Italian culture in the city. Don Francisco had made arrangements to book all the Italian artists for the summer season at Castle Garden, where their performances were enthusiastically acclaimed. And if Antonio wanted to, he certainly could continue to work as a theatrical engineer, with an introduction from Don Francisco, while conducting experiments in his free time to perfect his major inventions.

On Tuesday, April 23, 1850, the Meuccis bravely left Cuba aboard the American ship *Norma*. Antonio tried everything he could to carry their young daughter's body to the U.S. to bury her there. He had spent a fortune in a protracted attempt to find a substance or an electrical process to preserve her body after death, but failed. He had even postponed his departure for almost a month. The captain of the *Norma* was inflexible about transporting Juanita due to her death from yellow fever; the whole crew and passengers would have ended up in quarantine. After this latest exorbitant expense—so much invested for nothing—Ester took the family finances into her hands once and for all.

They arrived in only one week, on Wednesday, May 1, in New York, a rapidly growing port destined to become a world metropolis. It was already the undisputed center of trade between the Old World and the New World, with a population of over 700,000, three times that of Havana. Half of its inhabitants were recent immigrants, mainly of Irish and German descent. At that time there was a limited number of Italians, about 3,000, some of whom were notable political exiles who had escaped after the 1821 and 1831 riots and the 1848–49 uprisings. In the early decades of the nineteenth century, the U.S. had been the only country to offer unconditional support to Italian exiles, affording them a secure refuge—unlike England, which granted asylum to Giuseppe Mazzini, but refused to aid General Giuseppe Garibaldi while he was in Gibraltar. Most of these exiles had been sentenced to death for taking part either in the Giovane Italia or the Carbonari uprisings. These sentences could be converted to either serving twenty years in the horrific Austrian prisons of Spielberg and Gradisca or receiving permanent exile to the U.S.

This is why New York had become the vital center of a new Risorgimento. Garibaldi arrived there to escape a death sentence for his latest attempt to liberate the Italian peninsula with the proclamation of the Roman Republic in 1849. New York was also important to him due to the presence of brave veterans of the Italian Legion, which he had founded in 1842 in Montevideo, Uruguay, who followed him to Rome in 1848–49 and later shared his forced exile. Young and ardent patriots such as Giovanni Morosini, a Venetian who became a multimillionaire banker, were also in New York at that time. They would follow Garibaldi back to the Italian peninsula in 1860 for the pivotal Expedition of the Thousand, which would finally overthrow the Bourbon Kingdom of the Two Sicilies, based in Naples, enabling the union of Southern Italy and Sicily with the north.

Most of these exiles were people of great conviction, like Felice Foresti, Giuseppe Mazzini's representative in the United States.[17] He became the uncontested leader of New York's Italian community, which had its own newspaper, *L'Eco d'Italia*, first a weekly newsletter and then a respectable paper. While most immigrants remained marginalized outsiders, Foresti had successfully integrated into New York academia and society. His cultural stature was such that he was soon offered the Chair of Italian Language and Culture at Columbia University (then Columbia College), previously held by Venetian poet Lorenzo da Ponte, author of thirty-six librettos, in particular for the three operas of his dear friend Wolfgang Amadeus Mozart: *The Marriage of Figaro*, *Don Giovanni*, and *Così fan tutte*. Along with other talented and respected patriots, Foresti gradually formed the core of the early Italian colony, keeping alive the flame of hope for a free, united Italian motherland.

Italian immigrants were enterprising farmers and peasants,

tradesmen looking for opportunities, artisans and craftsmen providing artistic embellishments to affluent New Yorkers.

New York was already home to the immensely wealthy and those struggling in grinding poverty. Next to the most elegant mansions of the very rich, along the broad avenues of Manhattan, the urban fabric included many tenement houses that gave shelter to immigrants from around the world in the most crowded, inadequate, and unhealthy conditions.

The Meuccis were welcomed to this complex metropolis by their dear friends, tenor Lorenzo Salvi and musician Domenico Mariani, who took them to Mario Rallo's boarding house, one of the best in New York, where they lived for the first five months. They became painfully aware of having landed in one of the most densely populated areas, full of contrasts, and decided to stay together with their friends.

They initially settled at Leonard Street, at the intersection of Broadway, where many of their countrymen, artists, craftsmen, and opera singers were already living. This was a very convenient address near the Castle Garden theatre, then the largest and most important opera house in New York, located in Battery Park, at the southern tip of Manhattan. Here foreign opera companies used to perform and many famous singers made their American debuts, including Salvi, who started performing that very summer and was paid a fortune. Between 1839 and 1842 he had distinguished himself at La Scala in Milan in two of Verdi's world premieres and felt at ease on both sides of the Atlantic. While their friend was engaged as a tenor in several operas, Antonio and Ester worked for a little while as mechanical engineer and costume designer.

Around the corner from Leonard Street was an area called Five Points, later known as Little Italy, where another Italy was thriving—one of derelicts and criminals. Antonio and Ester were immediately aware of the degradation of these fellow citizens. This was the den of Italian slavery, practiced with impunity by unscrupulous Italian bosses, largely overlooked by the local police. Many immigrants would be paid half the wages they should have earned, only because they were illiterate and did not understand a single word of English. The Mafia and Camorra were running this white slave trade and perpetuating crime, extortions, and thefts unpunished, discrediting the Italian community for years to come. At the same time in the 1850s, more political deportees and common criminals from Piedmont and other parts of the Italian peninsula were arriving.

In 1855, the State of New York decided to impose stricter controls on immigration and established the Emigrant Landing Depot and Offices in Battery Park to process newly arrived immigrants. As a year earlier, the Academy of Music had opened—a 4,000-seat, acoustically engineered grand opera house, then the largest in the world—so the Castle Garden theatre in Battery Park had closed. The Emigrant Landing Depot and Offices took up residence in the empty theatre.

This initiative proved very useful for honest immigrants looking for work in the U.S. In spite of police surveillance, however, Little Italy remained full of speculators and scoundrels armed with *stilettos*, hovering over the new arrivals like hawks, ready to lure them by offering all kinds of services. Mysterious deaths occurred and went regularly unpunished.

After witnessing these abuses every day, the Meuccis decided they didn't want to settle in Manhattan. They had seen several apartments, by themselves or with the help of local brokers. Most tenement houses in Little Italy were run-down, overcrowded, and in poor sanitary condition, often without any hot water or heating in the winter or appropriate ventilation in the summer.

Then, on July 30, 1850, as announced by the *New-York Evening Post, the New-York Daily Tribune*, and *The Sun*, they learned that Giuseppe Garibaldi, known as the Hero of Two Worlds, had arrived. He had set sail in Algiers with his inseparable aide-de-camp, Major Paolo Bovi Campeggi, and after thirty-three days on the *Waterloo*, finally landed on Staten Island via Liverpool, England. This was where the Quarantine Grounds of the Marine Hospital Service was located. It was connected to Lower Manhattan by the Staten Island Ferry.

After being sentenced to death and excommunicated by Pope Pius IX for the vain defense of the Roman Republic, confronted by the far superior French and papal army, Garibaldi was rejected as an exile by half of Europe. He was emotionally drained and physically exhausted. His pilgrimage through Europe and North Africa to escape from capture had left clear marks on his body and spirit. During most of his trip he suffered from severe rheumatic pain and his left arm was immobilized. He was still mourning the recent loss of his wife, Anita, who died in their grueling attempt to escape. His dark velvet suit showed his respect and his grief. Given the exceptional heat wave, the Pavilion Hotel on the New Brighton coast of Staten Island, with its shady forests, turned out to be the ideal landing place to rest for a few days, a peaceful oasis not far from the busy life of Manhattan.

Among the first people to welcome Garibaldi was General José Antonio Páez, also exiled from his country, who had just arrived from Venezuela after fighting for its liberation from Spanish rule. At the Pavilion Hotel they had a long conversation in Spanish, with Garibaldi lying on a comfortable sofa to ease his pain.

Meucci shared the privilege of greeting Garibaldi, as he was part of a small delegation of distinguished Italian patriots that included, among others, General Giuseppe Avezzana, who had also been sentenced to death during the 1821 Piedmont insurrection. He had emigrated to New York and was welcomed as a hero after the 1849 siege of Rome.

Despite the beauty and the luxury of the Pavilion Hotel, Garibaldi was eager to leave and start the appropriate treatments to his left arm as soon as possible with his American physician, Dr. Valentine Mott Jr., who had fought in the Palermo uprising in 1848.

Garibaldi was forty-three and a world celebrity for his military exploits in South America, where he had fought for the independence of Uruguay, becoming the hero of Montevideo. His brave but ultimately disastrous attempt in 1849 to free Rome from papal rule brought him even more fame. Though a living legend, Garibaldi was very easygoing and reluctant to accept honors and compliments. He was always dressed in a red shirt with a black handkerchief around his neck to cover a wound suffered in San Antonio, Uruguay. He was tall and slender, with long reddish-blonde hair and a thick beard. His penetrating eyes and compelling voice were his most striking features. His priority during his forced exile was to recover his strength and plan a favorable occasion to redeem his country from oppression. He had nothing against religion, but he had fought against the power of the papacy because it represented an obstacle to the unity of Italy.

The tributes to Garibaldi included a parade down Broadway to Astor House, a luxurious hotel where a banquet in his honor with Mayor Caleb Smith Woodhull and the Italian community had been planned for August 3, then postponed to August 10 to accommodate his health. It was all organized by a special committee, headed by the leader of the New York Italian colony, Felice Foresti, with the participation of the newly arrived Antonio Meucci, patriot and writer Quirico Filopanti, businessman Michele Pastacaldi, General Avezzana, and Captain Antonio Figari, a merchant navy commander.

Garibaldi graciously declined the invitation from the Italian community for health reasons. His kind but firm letter to the Italian Committee was published by the *New-York Daily Tribune* and the *New York Evening Post* on August 8, 1850:

> *Gentlemen,*
> *I regret that I must tell you that my poor health continues and will make it impossible for me to take part in the public gathering that you are planning for Saturday next.*

The slowness of my convalescence and the uncertainty as to when I shall again be well make it impossible even to set a date on which I would be able to be with you in a manner worthy of your kind and flattering invitation.

I hope that you will allow me to repeat more strongly than ever my often expressed wish to see this planned gathering given up. No such public demonstration is necessary to prove to me the sympathy of my fellow citizens, of the American people, and of all true republicans for both the misadventure I have experienced and the cause I served.

Although the public expression of such feeling could be a great satisfaction for me, exiled as I am from my native land, separated from my children, lamenting the repression of liberty in my country through foreign influence, nevertheless, believe me, I would prefer to avoid it and instead, quietly and humbly become a citizen of this great republic of free men, in order to sail under its flag and pursue a career that would permit me to earn my bread and await a more favourable moment to liberate my country from its oppressors, foreign and non-foreign.

After the cause to which I have dedicated myself there is nothing closer to my heart than the approbation of this great people; and I am certain that I shall obtain it, when they are convinced that I have honestly and faithfully

> *served the cause of liberty, in which the American people themselves have given such a noble example to the world.*[18]

The letter won Garibaldi great credit for his determined character, sincerity, and modesty, adding to the acclaim he'd already gained for his courage in his military campaigns. He decided to leave the Pavilion Hotel to convalesce for fifteen days at the residence of General Carlo Ferrero, a valiant defender of the Roman Republic, who owned a villa some thirty miles north of Manhattan along the Hudson River. Garibaldi then moved into the mansion of a generous and patriotic importer of marble, Michele Pastacaldi, which had become the gathering point for the Italian exiles in New York. Felice Foresti, always an important contact for new arrivals, also lived here. In this same house, Meucci, along with a few friends, came up with the idea of sending $100 each to Garibaldi's old mother, known as Mamma Rosa, in Nice, France. At this time Meucci was considered a wealthy Italian inventor who, after fifteen gratifying years of work in Havana, was now in a position to help his less fortunate compatriots.

Garibaldi stayed at Pastacaldi's mansion for only six weeks. Once his health recovered, he accepted an invitation to Astor House on Broadway, close to where Meucci was staying. He had liked the Florentine inventor right away because he was a man of few words and felt trustworthy.

Garibaldi had many friends in Manhattan—in particular, Magnasco, the pseudonym used by Giacomo Brizzolara, the notably generous owner of a trattoria-grocery store at 19 Spruce

Street, where Garibaldi would meet General Giuseppe Avezzana, opera singers, tradesmen of Italian origin, as well as American and Italian literati almost every day.[19]

Both the Meuccis and Garibaldi looked forward to leaving downtown Manhattan and finding a quieter place to settle. Along with Lorenzo Salvi, they looked at a few apartments in Harlem as well as on Long Island. They were about to give up their search when, in early October 1850, they found the perfect home, thanks to the generous offer of Max Maretzek, the Castle Garden impresario, who was a good friend of Salvi's. Maretzek was about to leave on a long tour of Central America, and since he had rented two cottages on Staten Island as vacation retreats, he proposed keeping one for himself and subletting the other, located in the Clifton neighborhood, to the Meuccis. It was only a ten-minute walk from the dock.

Staten Island was in many ways a special place. Long home to Native American tribes, it was first discovered by Europeans in 1524, when Italian explorer Giovanni da Verrazzano sailed into New York Harbor. By the nineteenth century, although agrarian and maritime industries remained, resort and manufacturing industries were burgeoning, as was suburban development. When the Meuccis moved there, the island had become a paradise for tourists and a summer vacation spot for New Yorkers, with extensive woods and lakes where it was possible to hunt and fish. Easily reachable by ferry, it attracted a community of artists, singers, painters, and literati looking for an accessible retreat from the frenzy of Manhattan. Staten Island's population consisted mainly of Irish and German immigrants in the North Shore areas and some fourteen Italian families who first settled in Rosebank, South Beach, and later Port Richmond.

Antonio signed the contract with the proprietors in his own name and he and Ester moved in, along with Garibaldi and Salvi, in the fall of 1850. On October 21, Felice Foresti wrote a letter to Giuseppe Mazzini about Garibaldi, informing him that "the General has just gone to Staten Island to live with a certain Meucci in order to rest in complete freedom and live economically."

The house on Staten Island was named Forest Cottage. It was located west of Forest Street between Maple Avenue and Willow Avenue in the middle of a wood with many tall plane trees, even taller than the simple cottage where the Meuccis would live for the rest of their lives. Though the two-story house was modest, Antonio and Ester thought it was very special when compared to the tenement apartments in Little Italy, where most New York Italians were living. The location was exceptional. The lush vegetation became sparser as it sloped toward the ocean. The view from the attic windows was stunning, encompassing much of the New York Bay, the skyline of Manhattan on the west, and the nearby Brooklyn on the east.

Ester's dream had finally come true. She was happy, even though she had to take care of five men: her husband, their two friends, plus Garibaldi's bodyguard and Salvi's assistant.

The white clapboard house consisted of long front and back porches, a comfortable kitchen and a large parlor on the first floor, four bedrooms on the second floor, and a spacious basement, ideal for Antonio to set up a lab where he could place his entire arsenal of batteries and reels of insulated copper.

Garibaldi wholeheartedly accepted the Meuccis as friends and patrons. He and Antonio were the same age, both men of action, hardworking, and generous. The general was then an exile from his native land, without any income and unemployed. He did not have a place to stay, unless he accepted the hospitality of his wealthy sponsors in Manhattan, who enjoyed visibility and an intense social life. Garibaldi, suffering from a severe form of arthritis, needed a quiet place to rest and fully regain his health. Staten Island was the ideal private spot for him, but he intended to earn his keep rather than be a guest.

He knew he had to spend the next five years in exile, far from his mother and children, while looking for a way to obtain a ship to go back and resume the fight for the independence and unity of Italy. He could not support himself as a seafarer, at least not until his friend Francesco Carpanetto arrived in New York in April 1851. Together they would travel on Carpanetto's ship to Peru where, in Lima, Garibaldi was sure he could go back to the profession of his youth.

Meanwhile, Antonio's and Garibaldi's lives became unexpectedly and positively intertwined for eight months. They were motivated by the same deep love for their homeland, unfortunately enslaved to foreign rule, and entirely on the same wavelength. Both Antonio and Ester were struck by Garibaldi's honesty and discreet modesty, simplicity, and straightforwardness. His personality was magnetic, inspiring trust at first glance.

Garibaldi was very touched to be given the room with the ocean view. It made him think of his happy youth in Nice and his time spent at sea. In the next room was his bodyguard, Major Paolo Bovi Campeggi, who had lost his right hand defending the Roman Republic, and at night watched over the general's sleep.

Then there was a double bedroom where Salvi loved to stay when not singing in Manhattan, along with Domenico Mariani, his secretary and administrator, who would often cook for everyone in the house. Over the years he would become a devoted member of the family.

The master bedroom, where Ester and Antonio slept, was nearest to the staircase leading to the main floor. Here, in this room, the first permanent telephone line in the world would be installed in 1852, running from the bedroom window to the lab downstairs.

Antonio had arranged the ample basement as his laboratory, and equipped it with coils, wire, and orderly rows of batteries. In the first few weeks, while Garibaldi honored him with his presence, he set aside his most demanding experiments with the telephone to make a few pieces of furniture along with the general, using twigs, roots, and branches found on the island. Together they also made an experimental rocking chair, which turned out to be an artistic prototype, using wood that had been bent and steamed into the desired shape, without the need for a single nail.

In the meantime, Ester was busy sewing curtains and sheets for Garibaldi's room with the help of a sewing machine she had been given before leaving Havana. She gradually surrounded herself with pets: first, a cat, which was a gift from Mariani, then several more. Over time, it became a colony.

Garibaldi had come to Clifton with his inseparable companion: a macaw in a cage. It was unable to fly but could climb branches. He placed it on a trestle in the parlor. The macaw had been trained to say, "*Viva l'Italia, fuori lo straniero*!" ("Long live Italy, out with the foreigner!"). As soon as a visitor entered, the patriotic feathered bird would remind everybody of their true mission.

Besides his macaw, Garibaldi brought few belongings with him. His room—in addition to an iron bed, a small table, three chairs, a dresser, and a washstand Maretzek had left behind—contained only his traveling bag. It held a few pieces of clothing, one of which was the bright red shirt he wore during the defense of the Roman Republic; a dagger with a silver handle, chiseled and oxidized; two pistols; a small mirror; and a pair of scissors to trim his beard and hair, which he wore long to conceal an ear that had been severed during a campaign in South America. On the dresser stood a wax skull to remind him of the painful death of his beloved wife, Anita, who died of fatigue near Ravenna while trying to escape their enemies after unexpectedly joining him in the siege of Rome, pregnant and very ill. A memento mori that helped him put everything in perspective.

On Staten Island, Garibaldi's longing for the sea became clearer every day. He liked to go fishing and duck hunting. Antonio was happy to satisfy the desires of his illustrious guest, gladly buying a few rifles and a small single-mast sailboat that he painted red, white, and green with a matching sail. Garibaldi called it *Ugo Bassi*, in honor of the heroic Catholic chaplain and patriot who had joined him in the siege of Rome and was then abandoned by the clergy into the hands of an Austrian firing squad.

While talking with Antonio about his future in Staten Island, Garibaldi suggested that he create opportunities for jobless Italian exiles. This way he could protect them from the Mafia and the Camorra by encouraging them to live in the countryside, where life was less expensive, yet still close to Manhattan.

Among the various possibilities, Garibaldi's bodyguard, Bovi Campeggi, suggested experimenting with the production of sausages and salami, since he was from Bologna and a real expert in the field. At that time, this culinary specialty was very popular and was even sold on the Staten Island ferry. It was nearly winter, the right time to have a pig killed, possibly before Christmas, and then they could let the salami cure for a few months hanging from the basement ceiling.

Garibaldi, however, strongly recommended opening a candle factory, given the professional knowledge Antonio had gained while working at La Pergola and Teatro Tacón. This suggestion was also inspired by the significant amount of manual labor it would require, making it was an ideal project to provide employment to a number of Italian exiles in the area.

Investing in a candle factory also seemed a good financial idea because the price of tallow was dropping and vegetable oil from African forests was becoming increasingly more available and less expensive, yielding better results. Moreover, candles with these new characteristics could be successfully exported to Cuba and other locations.

The candle-making industry was practically nonexistent in the U.S. in 1850. Following Garibaldi's suggestion, Meucci became the first to manufacture stearin as well as paraffin candles in the American continent, devising a formula that he later patented in 1859. Unlike imported stearin candles, subject to heavy customs duties, the candles Antonio created were affordable and innovative, since they didn't leave any trace of smoke or odor. His unique invention consisted of manufacturing "mold candles mainly of paraffin and wax," which, up to that time, had been formed in molds of "impervious metal." He replaced these with molds of porous material. To lubricate them before use, he saturated them with a mixture of stearin with spermaceti—the fat of sperm whales—and/or natural wax. He obtained the best results by constructing molds out of plaster of paris that he would later successfully patent.[20]

Antonio spoke at length with Salvi and his assistant, Mariani, about Garibaldi's idea of a candle factory. Since he and Salvi were both in good financial shape, Salvi would buy the land and Antonio would buy the machinery. They agreed that for the time being, they could start making candles in the basement. Although the necessary equipment was very expensive, Antonio was ready to take the risk and buy it right away, before Salvi's return from Cuba for a singing engagement in late February 1851.

Antonio's candle factory required a large investment of capital to get started. First, there was the acquisition of equipment: two copper boilers, a steam generator, and other tools as well as an ongoing supply of raw materials such as tallow and wicks. Then, a number of workers were necessary, both skilled and unskilled. The tallow arrived in barrels that were to be picked up twice a week at the landing in Clifton. Skilled workers were required to supervise the complex processes of the fusion of the tallow, saponification, and crystallization into cakes. This was followed by the clarification of the stearin, which was finally placed into tubs and then into modeling machines, where reels of wicks had been previously inserted. The whole operation lasted about a week and a half in an area rigorously protected from air currents.

The candle factory was located in Antonio's basement. When Salvi came back from Havana, he became an official partner. On December 10, 1851, the two started buying land from Susan and William Townsend, the owners of Forest Cottage and part owners of the *New York Evening Express*. By mid-1853, they had bought the whole block for a total of $10,000.[21] The factory was soon built nearby, with the boiling vat located on the lawn outside because it was foul-smelling, large, and dangerous, as paraffin is highly flammable.

Salvi hired several expats, among them Giovanni Morosini, a brave and gifted Venetian young man who was easily recognizable because of a wound on his forehead, inflicted by the Austrians during the defense of his hometown in 1849. He had started working in Manhattan as a shoeshine boy and then came for a few months to the candle factory just to be close to Garibaldi, his idol.[22]

From the very beginning the factory was fueled by patriotic enthusiasm, producing candles in three colors—red, white, and green. Visitors would buy them as a tribute to Garibaldi, who represented a unique beacon of liberty. In no time the nearby cottage became the house of Italian exiles of all classes and a meeting place for political refugees of different nationalities.

The island was full of wildlife, and Garibaldi, along with Antonio, when he was free, would go hunting and fishing. They baked their own bread and ate the game and fish they had caught themselves. They were practically living with what the surrounding rural area could provide. Garibaldi liked to make his own cigarettes, rolling dry tobacco leaf in the South American fashion and, since he was a heavy smoker like Antonio, storing some in the chiseled wooden boxes he had crafted at night as a pastime.

Besides designing and building a few pieces of furniture, they were even able to build a celesta, a sort of fortepiano producing a kind of unworldly music, thanks to its quartz crystal bars that, once tuned, never needed to be tuned again. Since 1851 there was always at least one keyboard in the house because Antonio had learned to play quite well in his days at the Accademia in Florence.

In his memoirs, Garibaldi wrote:

> *My boss Meucci did not treat me as one of his working men, but as one of the family and with much affection, as Signora Ester did. At their home there was no luxury, although nothing was lacking in terms of food or lodging.*

Garibaldi used to speak of Antonio as "a friend of mine, a Florentine and a worthy fellow." They had a special relationship of trust and respect. While sailing he would call Antonio Capitan Buontempo (Good-Weather Captain) for his good-natured temperament, since he was always smiling. At work, in the candle factory, he would respectfully call him "Boss," because Antonio was the owner of the factory.

The candle business was initially a great success. It gave work to quite a number of exiles and Garibaldi was happy to give a hand to his hosts. Bovi Campeggi, on his part, was particularly skilled at finding commercial outlets for the merchandise. Antonio, however, was preoccupied with his various inventions, while the general was impatient to return to fighting in Italy and Salvi preferred to spend his time at the piano rehearsing arias for his next performance. How could such a factory possibly prosper in the long run?

Garibaldi set off on trips to South America, China, and Australia. Once he came back to New York in September 1853, he was met with the news that his mother, Mamma Rosa, had died six months earlier. He wanted to go home for a final attempt to liberate the Italian Peninsula, so he spent the next four months in New York trying to obtain a ship that he could captain. He stayed with Antonio for a few weeks prior to his final departure on the *Commonwealth* out of Baltimore on January 10, 1854.

During this time, Salvi decided to dissolve his partnership with Antonio, entirely ignoring the money Antonio had invested in their startup. It was not yet profitable, but in a year or two it definitely could have provided a fair income. Garibaldi intervened on Antonio's behalf, and Salvi finally recognized his debts to Ester and agreed to sell his shares for a symbolic amount. Still, he made them wait for almost a year before finally signing. Garibaldi made sure that Antonio and Ester were left with a valuable property and, as a further sign of appreciation, donated to them the iconic red shirt he had worn during the defense of Rome, along with a few other personal gifts. He also knew that Antonio had wisely intended to apply for American citizenship.

Upon Garibaldi's departure, Antonio immediately started thinking of closing the candle factory. First, Ester's health had further deteriorated since a major attack of gout, so not only was he unable to count on her support, but he also had to come up with ways to pay for her medical expenses, which included occasional high doses of morphine. Second, the startup was not yet making a profit, and kerosene had just been introduced to the market, resulting in a falling demand for stearin or paraffin candles. A few months later, Procter & Gamble started producing stearic candles in the U.S. and their profits skyrocketed.

Although Antonio was well aware he had been the first in the world to introduce the modern paraffin candle process, he realized there was no way he could compete with such a commercial giant. He knew he was a real innovator, even in candle manufacturing. Unfortunately, however, his partners had left. He was alone in a foreign country without any practical experience in financial and tax matters.

While managing his business and personal responsibilities, Antonio had installed the prototype of the *telettrofono*, his speaking telegraph made in Cuba, on Ester's night table. He powered it by Bunsen batteries, and added a bell that Ester could ring by pulling a cord, thus inventing a rudimentary ringtone. The very first telephone conversation ever held was between Ester and Antonio right there, at their home in Clifton. He had provisionally established two separate lines to prevent any echo.

Antonio had established the world's first telephone link. Their old telettrofono would someday become a household necessity.

Then, thanks to the intervention of Carlos Pader, a Spanish philanthropist he had met in Havana, Antonio gained access to Charles Chester, a manufacturer of telegraphic instruments in Manhattan, who enabled him to make significant improvements. They discussed the technical requirements in French, identifying and acquiring key parts, which allowed Antonio to make remarkable progress in transmitting the human voice through an electric wire and to construct a new prototype for a much more efficient telettrofono.

With this fresh supply of components, each experiment became increasingly satisfactory and rewarding. He now concentrated on the concept of a diaphragm that would make the human voice stronger and clearer and built the first of a long series of telephones made in the United States.[23] Each prototype consisted of "a tube of tin with the mouthpiece of the same material," according to Antonio's own description. "Beneath the mouthpiece there is an animal membrane saturated with bichromate of potash to render it firmer. In the interior is a small tongue of platinum soldered to a conductor of copper that communicates with the battery, and in the interior is a tube with wire all around it, forming something like a bobbin," he wrote. The animal membrane was a real step forward, a brilliant intuition. In the following years he raised the diaphragm to a position near the mouthpiece and introduced a bobbin or metal core with wires coiled around it. For the housing, he kept testing pasteboard and wood to see which functioned better.

The new phone incorporated three innovations to which Meucci dedicated an enormous amount of creative energy. The first was the bobbin, or core with coils; the second was the iron diaphragm; and the third was the small platinum tongue in the interior, soldered to the copper conductor connecting it to the battery.[24]

As he carefully recorded in his historically invaluable Memorandum Book, he developed about twenty telephone models from 1853 to 1860, steadily improving the quality of speech transmission. All of this thanks to his frequent shopping visits to Charles Chester in Manhattan and their enlightening conversations.

By the late 1850s Antonio had made a model that had all the requisites of a perfectly functional electric telephone. "It contained a diaphragm and a magnet with an electric coil around it. The voice made the diaphragm vibrate, alternating the current of the magnet. The alternations of current, being transmitted to the other end of the wire, caused analogous vibrations of the receiving diaphragm, thus reproducing the spoken word," he wrote.

Throughout Clifton the neighbors were aware that in the

Italian inventor's house, people spoke from one floor to another using a strange apparatus with wires. Many of them, as well as friends and acquaintances, asked if they could try out the unique speaking machine in the Meucci house, in particular Mary Gregory, a young woman who assisted the invalid Ester.

Ester's illness was one of Antonio's main motivations to make progress in this field. Her hands, usually busy sewing clothes for her neighbors, suddenly went numb. Within a year she was confined permanently to her bed and chair. She remained an invalid, suffering physical and psychological disorders for over thirty years until she died in 1884.

Antonio gradually improved the lines that permanently connected her bedroom with his lab in the basement, obtaining a steadily better sound. He developed the dry cell battery—nine years before Georges Leclanché's patent—and by August 1870 had obtained excellent results in the transmission of speech, for an extended range of one mile, using a cotton insulated copper braid as a conductor. Antonio called this latest 1870 model his "best telephone." A true success!

However, the path from his first American prototype in 1853–54 to the 1870 model proved to be full of obstacles, both financial and personal.

For the first five years the Meuccis, along with their friend Salvi, had acquired land in Staten Island. Though they could not count on any cash flow, Antonio and Ester had proudly become, for the first time in their lives, owners of the cottage in which they lived and of the surrounding lots of land, so that the candle factory could properly function and prosper. After Garibaldi's and Salvi's departures, the candle business rapidly failed. Creditors were already knocking at the door while debtors conveniently seemed to vanish.

Besides the constant research on his telephone, Antonio tenaciously tried to make a living constructing table fortepianos and celeste. The idea initially came from acquaintances in the opera world. Meucci thought they knew how to monitor the evolution of musical taste in the American population, but soon he came to the harsh conclusion that he would never receive any financial gratification in return. Over the years, building on his initial experiment with Garibaldi, he kept designing and assembling fortepianos with crystal bars with Salvi's former secretary Mariani, an accomplished musician who had become one of Antonio's good friends. It became Antonio's favorite winter pastime, especially because, on a small scale, it did not require additional workers or a significant investment in machinery. But the business was not successful.

Antonio wrote a letter to his brother Giuseppe, hospitalized in 1855 at the Ospedale Santa Maria Nuova in Florence, painfully sharing the details of his critical situation. He felt old and destitute; did not have the money to go back to Italy, not even for a visit; and the most terrible thing was that Ester was very ill and her health was not improving. She was suffering from rheumatoid arthritis, and after the general had left, her illness became chronic. Antonio's painful feelings of solitude and bewilderment are well expressed in this letter:

All I can tell you is that I am in very poor circumstances. Everything I owned has gone, and all I have left is the house and grounds and the candle factory. But it is idle to talk about candles because nobody wants them. I have now started making pianos, but this business is not doing well either. As you can see, I have no luck and everything I start seems doomed to failure. I wish I was living in poverty in my own country than here in this misery and fake freedom. This kind of life is for swindlers and thieves, not for gentlemen. In this country there is only selfishness. We are in the hands of foreigners making speculations on our skin and our lives.

Antonio was by now fully aware that he could no longer count on any help from his wife, previously a sage administrator and trustworthy financial advisor. He also realized he did not have enough experience to run a commercial activity on his own in a foreign country, too often ruled by self-interest, prevarication, and deception—particularly vis-à-vis immigrants, especially the ones who did not speak the language.

When Antonio had first arrived in New York in 1850, he had been considered a benefactor with a solid bank account, leading people to believe him much wealthier than he actually was. A very positive and optimistic person, he was a visionary well ahead of his time who patented about thirty inventions, from the lactometer to the hygrometer. Yet he did not manage to learn the language of the country where he chose to live, nor did he acquaint himself with American business law, regulations, and commercial customs.

He tried to find someone interested in acquiring his candle business because, though he still had his assets, he had no disposable income. By now he had spent most of his savings mainly to help the patriotic cause while giving work to as many Italians he could—to practically everybody who came to visit him asking for assistance. Unfortunately, due to his honest and trusting nature, in his new role of independent business owner whose partners had gone their way, he unwittingly opened the door to a series of impostors and ruthless speculators, always under the illusion they were his lifesavers.

On January 8, 1856, a man named Jim Mason, whom Antonio had met in Havana, came to his house and advised him to transform the candle factory into the first lager beer brewery in the U.S., proposing that they become partners. Mason was a flamboyant polyglot who flatteringly offered to personally take care of all the daily accounting and tedious administrative obligations, giving Antonio the freedom to dedicate all his time to new inventions.

Antonio liked the idea. Beer, especially lager beer, would be easier to sell than candles, and to brew the beer, he could use most of the general equipment and machinery he already owned, from the copper boiler to the many vats. Even the press could be used for filtering and separating the yeast. That was an encouraging start, he thought.

Beer required large quantities of pure water and the source available at Forest Cottage was the best on Staten Island, where there was not yet a brewery. Given the island's growing population of 20,000 inhabitants, it definitely seemed possible to supply beer to most of the shops and bars of the area.

What Antonio didn't realize was that he'd gained a very unreliable, deceitful partner—and from that very day, Antonio's trustworthy nature would condemn him to perpetual poverty. A rushed verbal agreement and a confident handshake between the two men marked the end of Antonio's financial independence, which would lead to a number of catastrophic legal consequences for himself and his wife. He naively trusted Mason and started signing not only promissory notes for large sums—to be paid to people he had never seen or heard of, on the basis of good faith—but also multiple mortgages on his property, which irrevocably compromised his finances.

In 1857 Mason managed to have one lot of the Meucci estate deviously assigned to him, with the pretext of having to pay suppliers and salaries for the new brewery. He became owner of part of the land. Ester, in spite of her illness and illiteracy, knew her mathematics very well and managed to have the remaining lots transferred under her name, including buildings, fixtures, and furniture, since they had been bought with her own money. During the same year Mason convinced Antonio to sign a ruinous partnership agreement, giving Mason full power over all the transactions of the brewery. The agreement downgraded Antonio to the position of a mere worker, leaving him a powerless owner. Within a few months of partnership Antonio became a pauper, swindled and abused.

At the very beginning, the Clifton Brewery ran smoothly,

even exporting beer to Cuba. In fact, it gave rise to what became the large Bachmann's Brewery, which, years later, in 1876, would win three gold medals at the Philadelphia Exposition, celebrating the centennial of the United States.

From the opening of the Clifton Brewery in 1856 until 1860, thanks to Mason's constant cheating, Antonio accumulated nine pending injunctions from him and his cronies. Antonio's defense lawyer, Henry H. Morange from New York City, proved to be both completely ineffective and financially greedy. Morange dragged the case out for over five years, finally closing it in April 1862. Because of the accumulated debts, he suggested that the Meuccis transfer all their assets to a man named Captain John Bennis in exchange for $8,501, but no buyback was negotiated and every agreement was merely verbal.

After this transfer the Meuccis indeed became dispossessed and would be forever indigent. Though many creditors were charged with fraudulent transfers of property, the situation was so mismanaged that in the end Morange himself sued the Meuccis to claim his legal fee. As a consequence, their house was auctioned off and Antonio and Ester saw their whole property, acquired with the savings from fifteen years of hard work in Havana, vanish within ten years of their arrival in New York.

While in Havana the similarities between the Spanish and Italian languages allowed the Meuccis to mingle with people and positively interact, in the United States they found communicating in English nearly impossible, except through an interpreter. Moreover, because of his nationality, Antonio was heavily discriminated against, even by American professionals.

Back then, Italians were often marginalized, as they were automatically associated with the Mafia or other criminal associations. They were not treated fairly in business and weren't permitted to hold management positions at most job sites. This made it very challenging for immigrants. All of this was happening in a big city like New York, where even smart and well-informed people could easily be swindled.

Due to his utter commercial naiveté, Antonio continued to meet a number of profiteers, and this led to his inevitable financial ruin. Ester was now literally paralyzed and in need of constant attention and care. Antonio, with his positive temperament and attitude, managed to transform even this calamity into a valuable opportunity by constantly testing his main invention. Whenever everything seemed lost, he rushed to his laboratory, not so much to forget his bad luck or out of pure irresponsibility, but because he placed his unwavering trust in the fact that the solution to all of their problems could be found in his promising telettrofono. He put the many disasters he suffered in the background, giving his wife the hope that she would soon be able to spend a peaceful old age.

Ester, however, had become more and more impatient with Antonio's fixation on inventions. She was tired of her husband shutting himself up in his laboratory as soon as he had a bit of free time. She could not tolerate his spending all that money on books and strange instruments that in her view served no real purpose.

During that troublesome time Antonio rigged up a system of practical communications that spared him from constantly climbing the steep stairs to the second floor, allowing him to speak directly with Ester as needed—not just in case of emergencies, but any time throughout the day because of his deep love for her. He soon extended the initial telephone line to Garibaldi's former room, which Antonio now occupied, to the lab in the basement and the external factory.

Ester found this new setup to be of great use and comfort. She used it to communicate with her husband, her caregiver, and with Domenico Mariani, their best friend, who lived with them for long periods when he could not find an engagement with an orchestra. To repay their hospitality, Mariani gave Antonio a hand in the lab, tended to the household chores, and even cooked for them on and off for almost nine years.

In her disfigured condition, Ester preferred not to be seen by visitors. Her telephone conversations with Mariani were of great help, especially because of his cheerful personality, which always managed to raise her spirits and ease her isolation.

In spite of every hardship, Antonio kept working relentlessly, improving the diaphragm of his telettrofono, which had evolved from animal membrane or parchment to metal. By 1856–57, his prototype had all the requisites of an electric telephone, becoming increasingly practical. The version built between 1858 and 1860 assured even better results.

In 1857, Nestore Corradi, a well-known miniaturist and painter, acquainted with Antonio for years and now living in New York, drew a sketch of the invention. It represents two seated men, each holding in their hands two small apparatuses of concave form, attached to electric wires. One man has placed an apparatus at his mouth in order to speak into it, while the other person has placed the other by the ear in order to listen to the human voice being transmitted through the electric wire. Antonio himself commissioned this drawing of his unique invention, named Telephone 21, in the hope of finding sponsors who would believe in him. All they had to do was see the drawing, trust his unique technical knowledge, and come to his house.

Unfortunately, he lacked the vocabulary and the specific terminology to explain this landmark invention, so far ahead of its time. He also lacked the political connections or sufficient funds to assert his own rights as the inventor and bring it successfully to the market. This drawing was made nearly twenty years before Alexander Graham Bell pretended to be the first to speak across a wire, in 1876. So the real question was: How could an immigrant with no money promote this monumental invention that would change interpersonal communications around the world?

Given his linguistic limitations, Antonio thought that if anyone might be willing to help him, it would be his own countrymen. He begged some rich Italian families for assistance, but did not get very far. In 1860, *L'Eco d'Italia*, the only Italian newspaper in New York, published a leading article on his latest invention. Antonio was hoping this would stir some interest. It didn't help.

Crushed but not beaten, Antonio relentlessly kept working on the telephone. Having come to America firmly believing it was the land of opportunity, he totally trusted the potential of his invention. He thought that by perfecting the design and finding the right financial backers, his hardships would pay off. He was possessed by his inner conviction, though Ester would in private reproach him for dedicating too much time to any invention requiring such a substantial investment with no financial return. She was, however, equally ready to defend him in public, even with Father Gioacchino Parodi, their parish priest, who would often visit them.

During this time in 1858, Antonio was introduced by a friend he had known since Havana—the retired tenor Domenico Lorini—to William E. Rider, who had married the sister of Lorini's wife. Rider was from a very wealthy family. His father owned half of the Goodyear patents for rubber production, which provided a very high income. Rider immediately understood that Antonio's multiple inventions could turn into a goldmine for him and decided right away to undertake some profitable venture of his own.

He was soft-spoken, kind, and elegantly dressed, and smoked expensive cigars. He introduced himself as an enlightened patron, and with his upright and gentlemanly appearance he easily managed to impress the impoverished Italian couple. The Meuccis did not realize he was a shrewd speculator and mistook him for a saving grace. At that time Meuccis' property was heavily mortgaged and the judicial proceedings and lawsuits were still pending. Rider was a real enticer and allurer, ready to take advantage of the Meuccis' precarious situation and creativity for his own interests, gradually revealing himself as a financial predator and a ruthless profiteer. By expressing great admiration for Antonio's multifaceted inventiveness, Rider gradually gained his complete trust, and even more so, Ester's confidence. Later on, she naively entrusted him with over $11,000 (more than $122,000 in today's dollars) to safeguard in case of emergency, without informing her husband. This large sum was never returned.

The first deceitful thing Rider did was offer to accompany Meucci on a trip to Chester's New York shop to buy anything he needed. In spite of his financial and personal troubles, Antonio continued to believe in the validity of his main invention. Ignoring his creditors, he persistently kept buying batteries, magnets, and whole bundles of copper wires for at least $100 each time. Antonio was known as a very honest inventor, gifted with an irrepressible creativity, and often bought everything he needed on credit. This time, in August 1858, thanks to Rider's lavish offer, he could finally stock up on electrical material and batteries without worrying about how to pay for it. He was as happy as a kid.

Whenever Antonio came back to Clifton, some viewed him with suspicion, while a crowd of young admirers enthusiastically waited for him to further test his inventions. Over the years his cottage became known as The Devil's House, and Antonio was seen by some locals as a person rich in mysterious knowledge, a magician, while they considered Ester a faithful muse or saint.

Ironically, it was thanks to Rider that the apparatus he constructed between 1858 and 1860 turned out to be such a fine electric telephone. It was the first he made with the bobbin he bought from Chester, Antonio wrote, "with its center made with a piece of steel, tempered and magnetized permanently, put inside a circular box of pasteboard with a wooden bottom. Above the bobbin was an animal diaphragm with a hole in the middle that covered a metallic tongue of iron, which served as a valve, opening and closing with the oscillations of the voice." The instrument performed very well; it was the immediate result of the new bobbin and all the other materials bought at that specialty shop.

With access to such a variety of supplies, other visionary projects moved ahead. Antonio managed to make a battery that could move a locomotive, this way envisaging electric traction for trains and industrial machinery decades before railway electrification became mainstream. Rider, however, was not competent enough to understand these technical matters and their potential applications in depth. He embraced down-to-earth projects, such as the invention of a special metal mold for producing clean, polished, and perfectly shaped candles, as well as a fiber wick that gave a far better light than any other candle before. He shrewdly offered to pay for the patent for both inventions—of mold and wick—and proposed that he and Antonio become partners in the candle business.[25]

They signed a contract by which Rider would provide the money to develop Antonio's inventions. The agreement stated that the patents were to belong to Rider, who would keep the right to select and pay only for those inventions that he deemed advantageous while they would share the profits. The first patent application, for a special plaster mold to produce paraffin candles, was filed in November 1858 and granted toward the end of January 1859, with the number 22,739. This was the first in a long series of patents that Antonio applied for and obtained in the U.S.

Antonio had an idea for a new candle factory, capitalizing on the experience he'd acquired with the stearin candle factory previously set up with Garibaldi, in partnership with Salvi. The New York Paraffine Candle Co. opened in February 1859 on Amos Street, close to the Vanderbilt Landing in Clifton. It was a real success, but Rider made his naive Italian partner believe that initially there was no profit and he kept most of the net income. The factory burned down about a year later and Rider managed to collect $11,000 from his fire insurance policy. He did not give one cent to Antonio, using the excuse that he had to invest the whole amount for another factory in Stapleton—a small neighborhood close to Clifton—where he rented and later purchased another building.

When the fire destroyed the whole building in Clifton, several articles appeared on the topic. When the new nearby factory was opened a few months later, a special feature was published in *The Richmond County Gazette* on Wednesday, December 12, 1860, underlining the complexity of the whole process of making paraffin candles as well as Antonio's pleasing manner and unique sense of hospitality:

> *Under the gentlemanly conduct of Mr. Meucci, who has charge of the works, we made an interesting survey of the premises, and were surprised at the number of distinct operations which must be gone through with before a box*

of candles is made ready for the market. First, the material (which is the residue of coal oil after kerosene has been expressed from it) is melted down in immense vats or jars; from this it is taken out in dippers with long spouts, through which it is poured into perpendicular molds. The wick is already hanging in the mold, and the candle having hardened around it, is taken out and cooled in water. After this operation, it is carried to the floor above, where the wick being cut, and passing for this purpose through some half dozen female hands, it is subjected to the action of a rapidly revolving knife, which rounds it off and gives it a symmetrical appearance. After this, being scraped, and smoothed, and polished, it is declared finished, and sent to the packer, who stores it away in a box, in company with hundreds of others, to be sent off to market.

The New York Paraffine Candle Co. continued to operate in the new location until 1866. Antonio remained contractually linked to Rider until 1869. According to a devious clause in the contract, Antonio was entitled to receive only 5 percent of the profits of the candle factory. However, as the losses in the first year amounted to $25,000 and no profits were declared until the end of 1863, Antonio had to live on his salary as factory manager, $20 a week, soon reduced to $15.

Such a salary would have been utterly humiliating for the superintendent of a factory that produced 1,000 candles a day, with a turnover of about $2,800 a week, as Antonio specified in several letters to his brother from 1859 to 1864. Antonio was only fifty-one and already felt old. Bitterly realizing he had made a speculator such as Rider rich, in 1859 he complained to his brother in utter sincerity by admitting that he had "sold and pawned everything," including his unique inventions. He "did not have enough to buy a piece of bread" and was "compelled to work at only $15 per week, a pay that here is given to a porter, a mere laborer."

At the same time, he decided to inform Garibaldi of his unfortunate financial downfall. He framed the issue in terms of a national historical challenge of living abroad in a hostile environment, complaining that he had been "living in misery for the last three years, since 1857."[26] He advised that he was sending him a case of white candles, with some red and green as well, along with a very precious gift: a rifle cleverly hidden underneath the false bottom of the candle crate. He was reconfirming his solidarity to the "Hero of Two Worlds," who in 1860 would finally engage with the Expedition of the Thousand to bravely unite the Italian peninsula, then still fragmented under foreign rule.

He affirmed in writing his deepest love for a united Italy: "*Amico Generale, altro non desidero, e dopo, la morte*" (My dear Friend General, nothing else beyond this I desire, and after this, only death). If Antonio and Ester Meucci could have afforded it, they would have returned to their native city, Florence, now part of Italy by popular vote since March 22, 1860, and later to even become the capital of the country, as of June 1865. He concluded his deeply felt letter by stating, "I hope to see Italy before I die and embrace you all . . . my misfortune as well as that of all of us, Italians, is that of being in the hands of foreigners in the New and in the Old World."

Antonio's hopes were revived when one of his closest friends, the coral dealer Enrico Bandelari, decided to visit Italy in 1860. Antonio had made many experiments for him, trying to turn red coral into the more valuable pink. Before leaving for Europe, Bandelari offered to promote Antonio's invention of the telettrofono. Antonio gave him drawings, a technical data sheet describing how it worked, and possibly the detailed article that had appeared in *L'Eco d'Italia*, clearly specifying that it was an electrical instrument.

Bandelari landed in Liverpool and proceeded onward to Paris and Genoa. Unfortunately, the timing was hardly propitious, especially in his native Naples, where he had hoped to enlist the direct support of Garibaldi, who had just triumphantly put an end to the Bourbon realm. Garibaldi, however, was totally absorbed by his heroic mission, and despite his friendship with the Meuccis, was unable to help. The citizens of Naples had been called for the first time ever to decide in a plebiscite whether they wanted "Italy one and indivisible with Victor Emmanuel as king." The Neapolitans had important decisions to make.

After days of waiting, Bandelari was informed by the Ministry of Posts and Telegraphs that the proposed invention, the electrical transmission of the human word, was judged to be literally impossible. The Ministry was not interested in financing this kind of utopian project. Other inventors from France and Germany, such as Charles Bourseul and Johann Philipp Reis, had previously tried to promote the same or very similar communication systems at a distance through electricity, succeeding only in terms of sounds but not the articulate human word. Bandelari returned from Europe without finding a single investor.

By then, Antonio, who had taken a ten-year mortgage of $3,000 with a heavy heart in 1856 from Augustus Samanos to pay the family bills, was confronted with the sudden death of the old gentleman. Samanos's son Stephen took over the mortgage in 1861, and with the complicity of the local sheriff, bought the whole estate for an insignificant amount of money by rigging the auction as the only participant, only to resell the estate immediately at a profit.

The Meuccis had now lost everything, but finally managed to get rid of all their debts. The only positive thing was that the new owner of their house, Frederick Bachmann, gave them the opportunity to stay in the house without paying rent in perpetuity. Ester had gradually been forced to sell most of the jewels she had been given in Cuba, so they ended up living off the generosity of their friends as well as small contributions of money from the government.

Despite their poverty, Antonio was still a partner in the New York Paraffine Candle Co., which even during the American Civil War did very well, finally providing some income from 1863 until 1866. The success of the new candle factory proved that Antonio's original idea in 1850 to set up a candle factory after his arrival in New York was not unreasonable. As soon as the factory's industrial production began, the price of paraffin was much lower than stearin and wax. Later on, he came up with a mixture of mineral waxes that was even more convenient than paraffin. For most of the time Antonio was in charge of the candle production in Clifton, with Rider supervising the sales in New York. This further allowed Rider to keep pocketing the profits undisturbed.

Rider never suffered financial troubles at any point throughout their business partnership. From 1858 to 1866, they submitted fourteen patent applications and two caveats—a document allowing inventors to notify the U.S. Patent Office of their intention to file a patent application in the future—nine of which were granted. Besides the patents for paraffin candles that changed the history of candle making, patents concerning oils to be used for paints and varnishes were granted in 1862 and in 1863, respectively. The latter was assigned to Ester, just to reinforce her trust.

Antonio kept researching various oils that would be of interest to ship and boat owners. He patented and produced siccative mineral oils and varnishes at his own expense, which were then shipped to Rider for sale. These products preserved wood against dampness and attacks of woodworm and insects. The oil in particular proved an excellent and cheap substitute for linseed oil, and could be used without turpentine to resist the action of air and water.

In those years Antonio continued to pursue challenging inventions. He documented everything in his Memorandum Book—a big, solidly bound, unused ledger Rider had given to him as a substitute for the pocket notebook he always kept on hand. Antonio methodically and meticulously noted every detail of his work, devoting great attention to the cost of materials and processes. He always tried to save as much as possible in the interest of his partner as well as the consumer.[27]

Among the many other profiteers who blatantly took advantage of Antonio's intrinsic honesty was Antonio Jané, a rich merchant from Cuba, who asked him to devise a kerosene oil lamp without using a glass chimney to sell in Havana. Jané solemnly agreed to pay the expenses of the patent and give Antonio half of the earnings, but once he had the registered patent in his hands, he disappeared, leaving no trace.

In a letter to his brother in 1863, referring to his new oil but also to the events of the Civil War, along with comments on the inflationary printing of currency, Antonio sarcastically suggested that his patent oil could also destroy certain other worms—enemies of Italy:

> *Here everyone steals, as all Governments do. The same that sentence thieves to jail. Those who come from wealthy families can do anything they want. If a poor man is in debt, he loses his honor. If the Government and the rich make debts, they are honored, you see what a fine world this is!*

On a much larger scale but on the same note, in 1864, President Abraham Lincoln, a year before he was murdered, wrote to Colonel William F. Elkins:

> *I see in the near future a crisis approaching that unnerves me and causes me to tremble for the safety of my country . . . Corporations have been enthroned and an era of corruption in high places will follow, and the money power of the country will endeavor to prolong its reign by working upon the prejudices of the people until all wealth is aggregated in a few hands and the Republic is destroyed. I feel at this moment more anxiety for the safety of my country than ever before, even in the midst of war.*

Fortunately, Antonio's inventive mind kept him alive and made him ignore all the obstacles he kept encountering. Many people came to ask him various things, and he always "served them with a good will, working for nothing," showing his great empathy and generosity toward others. Since the candle-making operation had proven successful and the telettrofono was fully functional at home, he gradually became busy in more challenging experiments, such as paper manufacturing. He began experimenting in new brave directions, such as making paper from wood pulp.

At that time, rags were the main ingredient in paper making. In the spring of 1864 at his Stapleton factory, he started to experimentally make paper without rags. Since large forests were available on the American continent, using wood as raw material for paper making was very attractive. Rather than grinding the wood into a sawdust-like powder, as previously attempted by other inventors, he managed to create a wood pulp rich enough in fibers and filaments to give it the right consistency and strength to make high-quality paper, which he successfully managed to patent. During the same year he patented a method to integrate the soda process, then in use with hydrochloric and nitric acid, to remove the gummy and resinous substances from the wood fibers. This produced a bleaching effect that lightened the color of the paper, making the written word more easily readable.

In 1865, he patented a way to reduce the fragility of the paper—which was typical of the soda process—by mixing a certain amount of cottonseed oil or olive oil into the caustic soda. With the third patent in 1866, he extended his method in order to render viable the processing of inferior vegetable material, such as straw, hay, tree leaves, Indian corn, weeds, old ropes, and more.[28]

Although nowadays 90 percent of the paper produced worldwide is obtained from wood, in 1864 Antonio's patents on paper-pulp making were visionary. They represented a radical innovation in the history of papermaking.

Rider must have been astonished when, in the fall of the same year, Antonio presented him with a sample of paper that was produced with his unique wood pulp, obtained from spruce wood, a type of tree that was and still is very common in large parts of North America. Impressed by the innovation, in particular the unexpected consistency, strength, and color of the paper, Rider immediately grasped that it could become an excellent, profitable business. He cleverly decided to contact the Associated Press in New York. Through them it would be possible to influence almost all of the agencies that operated in the printed paper sector. Rider did not have any trouble convincing David

Craig, general manager of the Associated Press, to enter into a formal contract, which was signed in 1865 and based on Antonio's three patents. Always looking for better paper on which to print his newspapers, Craig agreed on $100,000, payable at a rate of $10,000 a month.

Antonio was convinced that a turning point in his professional life had finally arrived and proudly informed Garibaldi of this good news. Instead, the deal soon turned into an impossible mission. Rider agreed to set up a paper processing plant in the state of Ohio, where the American Civil War was raging. Antonio and two workers from Clifton were sent for the preproduction stage to Steubenville, a remote town about 500 miles from New York City, far away from the invalid Ester. Rider kept the initial sum of $10,000 for himself, half of which should have been given to Antonio. After a few months the plant had to close down, mainly because the war bankrupted most investors in this sector. The long, bloody Civil War, which saw the assassination of President Abraham Lincoln in April 1865, ended in that same year, marking the victory of the Union over the Southern Confederates and the end of slavery, which threw the economy into a shambles. Antonio, after receiving no money except his board and expenses for setting up the plant, found himself adrift.

The shameless dishonesty of his partner was now obvious. Rider had finally revealed himself for what he really was: an unscrupulous and rapacious adventurer who had managed, more than anyone else, to exploit and cheat Antonio.

For over seven years Antonio was subjected to despicable work conditions and financial enslavement. It did not help that Ester was captivated by Rider's personality and ended up paying a high price for her credulous naiveté.

Antonio then worked for six months at another paper plant, owned by Rider himself, at Perth Amboy in New Jersey, where he could experiment extensively, and with success, to obtain pulp from a large number of annual grasses. While Rider decided to sell his New York Paraffine Candle Co. to derive capital for the new venture, Antonio agreed to manage the new factory, two hours away from Clifton. To get from his house to the new factory, he would take the train to Tottenville in Staten Island, and then the ferry to Perth Amboy. Sometimes he stayed at a hotel there for a night or two, or even for the whole week, often at his own expense.

During this time, Ester's rheumatoid arthritis was so advanced that she seldom left her room. She was practically immobile and Antonio had to hire a reliable helper, Mary Gregory, to be in the house starting in October 1865. Every day the three of them used the telettrofono to communicate with one another.

Only in 1869 was Antonio finally able to break his exploitative working relationship with Rider, who in the meantime had surreptitiously managed, undisturbed, to empty the Clifton warehouse, sell all the candles, and, as usual, keep the whole profit for himself. Meanwhile, Antonio never gave up his experiments with the telephone. On August 17, 1870, he succeeded in transmitting the human word at a distance of about one mile using a reel of cotton-insulated copper wire, and was very proud of this new step ahead.

Chapter Four

STATEN ISLAND, NEW YORK (1871–1889)

Throughout their lives in America, the Meuccis did not learn to identify predatory or dishonest behavior until it was too late. They never developed the necessary strategies to save themselves, except when they had already lost everything. On top of all of this, Antonio was the hapless victim of an unprecedented disaster on the Westfield Ferry connecting Manhattan to Staten Island.

Sunday, July 30, 1871, was a hot day, and over 400 people, including many women and children, were leaving Manhattan for a peaceful afternoon on Staten Island to breathe some pure sea air and enjoy the shade of a lush, reinvigorating natural environment. Antonio had just left Chester's shop in Center Street with a supply of batteries and cables and almost missed the boat. At the very last minute a sailor helped him climb on board. All of a sudden, the steam boiler exploded, thrusting most passengers into the air and back onto a heap of broken incandescent iron and into the water full of boiling oil, as reported in detail by the *Tribune*, *Harper's Weekly*, and *Frank Leslie's Illustrated Newspaper*.

About a hundred people died and over 200 were injured, mutilated, or severely burned. Only the pilot was miraculously uninjured. The terrible explosion was due to the boiler's flawed safety valve.

Antonio was drenched by the boiling water and thrown into the bay. His name was mentioned by the New York press, among the few personalities who had been identified as victims of this disaster, proving that he was indeed well known. He was taken, along with all the other casualties, to the Park Hospital. He was covered with ghastly wounds and burns on his whole body; his skin was almost entirely scalded, particularly on his face, neck, and chest, and he had lost large portions of his hair. When he arrived at the hospital, the doctors could not do anything except provide a few injections of morphine. His condition remained critical for some time, teetering between life and death. His family was notified, but after a week he unexpectedly opened his eyes, in a large ward where only male indigent immigrants were being hosted and treated. His fate continued to be uncertain.

Meanwhile, Ester, desperately in need of money, tried to take financial matters into her own hands. While Antonio was still at the hospital, covered with burns and on the verge of dying, Ester, with her caregiver Mary Gregory's help, sold all his telephone prototypes for a few dollars to John Fleming, a dealer in secondhand goods. It was an unfortunate decision that would later have crucial consequences.

After some time, the doctors decided it was more humane to send Antonio home for his final days. So Antonio, still in great pain, was taken back to Forest Cottage, where he stayed in bed in Garibaldi's old room for over three months. His German neighbor, Frederick Bachmann, the strong new owner of the beer factory, came every day to help him turn in bed. Mary Gregory would then disinfect the wounds, change the bandages, and put clean sheets on the bed.

As the Meuccis lived in utter poverty, they soon accumulated over $1,000 of debts for the medical expenses. The Staten Island Railway Ferry Company, responsible for the catastrophe, declared bankruptcy. Antonio tried to find a lawyer who could help him claim at least part of the damages, but he received nothing. The Staten Island Railway Ferry Company never paid any compensation to anyone.[29]

When struck by this disaster, Antonio had just left the Perth Amboy Fiber Paper Factory and was completely broke. He had been working most of the time without a salary and so far away from Clifton. After miraculously surviving the apocalyptic explosion, he was so poor that he was assisted by the Poor People Department and received $1 a week from the county, even though the treatments prescribed were extremely costly. A few friends and compatriots started collecting some money for his medical expenses, but that was far from enough. The Italian Consul General Ferdinando de Luca also helped him with other necessities, since Antonio had always proven to be an outstandingly honest, hardworking, and intelligent person. But Antonio's long illness without any security reduced the Meuccis to complete destitution.

After three long months convalescing in bed, Antonio won his struggle against death. After everyone had thought his time had come, he unexpectedly found the strength to descend the two flights of wooden stairs leading to the ground floor. He walked out of the front door onto the porch to take a deep breath of brisk November air. At that moment he felt the energy of his best years return and went down to his lab.

He found it empty. Dumbfounded and shocked, he could hardly believe that his beloved wife, with whom he had shared his life and dreams, could have disposed of all the prototypes he had constructed with his own hands during more than twenty years of work.

This loss was by far the greatest misfortune of his life.

As a result, he could hardly speak to Ester for days. Antonio, like the biblical Job, was nearly overcome by pain. How much could someone suffer and still remain loyal to his unwavering beliefs?

He did not let his rage take over. Instead, he patiently reviewed and updated the drawings and the documents he had given to Bandelari long ago as his friend left for Italy. After Antonio managed to contact the pawn shop owner who had reluctantly bought all his prototypes, he discovered that everything had been sold to an unidentified gentleman in New York City. He decided to take immediate action and do everything in his power to establish once and for all that these were *his* inventions.

He was consoled to find the drawing that his friend Nestore Corradi had made in 1857 showing two people using his telephone. From that moment on, all his energy was spent in asserting his ownership of this key invention.

He still felt betrayed and humiliated by the woman he loved and respected so much. He was also very worried that the unknown gentleman who bought all his prototypes could, and would, steal his project.

Recently he had come across a disturbing article reminding him that Innocenzo Manzetti, an inventor from Aosta, Italy, had come forward claiming that he had electrically transmitted the human word successfully. Antonio immediately wrote a detailed description of his invention, attached it to Corradi's drawing, and went to a patent lawyer, Thomas Stetson, accompanied by his friend, notary public Angelo Bertolino.

During Antonio's long convalescence, he and three of his Italian best friends formed the Telettrofono Company. The purpose of the company was to promote Meucci's invention as well as commercially develop the ability to transmit the human voice over longer distances.

His three associates were Sereno Breguglia, a tobacconist from New York; Angelo Zilio Grandi, secretary of the Italian Consulate in New York; and Antonio Tremeschin, a builder from Vicenza, Italy. They all met on December 12, 1871, at the popular, conveniently located Hoffman Cafe, in front of the Stock Exchange, where Breguglia had his cigar stand and luxury dining facilities were available. Over coffee, Bertolino, the notary public, drafted the agreement between Antonio and his three partners. They put together $20 and committed themselves to securing patents for this invention in America, Europe, and the rest of the world.

During Antonio and Bertolino's initial meeting with Stetson, the patent lawyer, they were informed that the patent application would cost $250 because it would require a great deal of time from the lawyer's staff to prepare an accurate and convincing description. As this amount was unaffordable for Antonio and his co-partners, Stetson encouraged them to file a caveat for $20, along with the documents Antonio had so carefully prepared.

A caveat, at that time and up until 1910, was a legal document issued by the U.S. Patent Office that provided an inventor with the exclusive right to turn his or her idea into a working, patentable invention for a period limited to one year. It afforded the same rights as a patent and allowed the applicant the exclusive opportunity to complete the invention.

The caveat was sent to the Patent Office in Washington and registered on December 28, 1871, with the number 3335 under the title "Sound Telegraph" and, according to the law, was meant to be kept confidential. This is the text of the caveat:

> *I employ the well-known conducting effect of continuous metallic conductors as a medium for sound, which increases the effect by electrically insulating both the conductor and*

the parties who are communicating. It forms a Speaking Telegraph, without the necessity for any hollow tube.

I claim that a portion or the whole of the effect may also be realized by a corresponding arrangement with a metallic tube. I believe that some metals will serve better than others, but propose to try all kinds of metals.

The system on which I propose to operate and calculate consists in isolating two persons, separated at considerable distance from each other, by placing them upon glass insulators; employing glass, for example, at the foot of the chair or bench on which each sits, and putting them in communication by means of a telegraph wire.

I believe it preferable to have the wire of larger area than that ordinarily employed in the electric telegraph, but will experiment on this. Each of these persons holds to his mouth an instrument analogous to a speaking trumpet, in which the word may easily be pronounced, and the sound concentrated upon the wire. Another instrument is also applied to the ears, in order to receive the voice of the opposite party.

All these, to wit, the mouth utensil and the ear instruments, communicate to the wire at a short distance from the persons. The ear utensils being of a convex form, like a clock glass, enclose the whole exterior part of the ear, and make it easy and comfortable for the operator. The object is to bring distinctly to the hearing the word of the person at the opposite end of the telegraph.

To call attention, the party at the other end of the line may be warned by an electric telegraph signal, or a series of them. The apparatus for this purpose, and the skill in operating it, need be much less than for the ordinary telegraphing.

When my sound telegraph is in operation, the parties should remain alone in their respective rooms, and every practicable precaution should be taken to have the surroundings perfectly quiet. The closed mouth utensil or trumpet, and the enclosing the persons also in a room alone, both tend to prevent undue publicity to the communication.

I think it will be easy, by these means, to prevent the communication being understood by any but the proper persons.

It may be found practicable to work with the person sending the message insulated, and with the person receiving it, in the free electrical communication with the ground. Or these conditions may possibly be reversed and still operate with some success.

Both the conductors or utensils for mouth and ears should

be, in fact I must say must be, metallic, and be so conditioned as to be good conductors of electricity.

I claim as my invention, and desire to have considered as such, for all the purposes of this Caveat.

The new invention herein set forth in all its details, combinations, and sub-combinations.

And more especially, I claim:

First. A continuous sound conductor electrically insulated.

Second. The same adapted for telegraphing by sound or for conversation between distant parties electrically insulated.

Third. The employment of a sound conductor, which is also an electrical conductor, as a means of communication by sound between distant points.

Fourth. The same in combination with provisions for electrically insulating the sending and receiving parties.

Fifth. The mouthpiece or speaking utensil in combination with an electrically insulating conductor.

Sixth. The ear utensils or receiving vessels adapted to apply upon the ears in combination with an electrically insulating sound conductor.

Seventh. The entire system, comprising the electrical and sound conductor, insulated and furnished with a mouthpiece and ear pieces at each end, adapted to serve as specified.

In testimony whereof, I have hereunto set my hand in presence of two subscribing witnesses.

ANTONIO MEUCCI
Subscribing witnesses:
Elias P. Lyman, William H. R. Reeve.

When Antonio's lawyer Stetson sent him the draft of the caveat for approval, Antonio found the description incomplete and added key explanatory notes in pencil in the margins. The lawyer, not entirely understanding the complexity of Antonio's invention, erased the clarifications and sent the original unannotated caveat to Washington.

When Antonio received the registered caveat, he immediately realized how many significant details had been omitted and how careless and hasty the lawyer had been. On January 12, 1872, Antonio wrote a formal letter to Stetson asking for an estimate of the costs required to describe his invention more accurately.

At the beginning of the same year, the newly founded Telettrofono Company encountered some major difficulties. Antonio Tremeschin moved back to Italy. Angelo Zilio Grandi, assigned elsewhere by the Italian government, withdrew from the partnership and sold his share to Sereno Breguglia, who passed away a few months later, leaving Antonio as the sole survivor.

During this painful period of recovery, Antonio was unable to work and found himself "reduced to little or nothing." He had great difficulty obtaining the necessities of life and the materials to continue his experiments. Nonetheless, Forest Cottage by now was equipped with three telephones that worked perfectly, and Antonio's strong determination and faith in his invention remained unshaken.

Before dying, his former partner Breguglia, who had proudly believed in Antonio's telephone project, managed to introduce him to Edward B. Grant, vice president of the American District Telegraph Company (ADT). Grant would regularly buy cigars at Breguglia's stand on his way to work at 62 Broadway, one or two blocks away from the Western Union headquarters. The offices of the Associated Press as well as those of Antonio's lawyers, the Lemni and Bertolino Law Firm, were all nearby. Breguglia had sincerely hoped Grant would turn into a major sponsor, possibly allowing Antonio to perform field trials of the telettrofono on ADT's telegraph lines.

In the summer of 1872, Grant warmly welcomed Antonio and his friend and mentor, notary public Angelo Bertolino, who acted as interpreter. He showed great interest in this intriguing proposal and asked to see the drawings and the detailed description right away, indicating that he would like to read it carefully at his own pace.

Antonio was then sixty-four, yet still very trusting and naive, and he was so intimidated by Grant's important position that he agreed to leave all the papers there, as advised by Bertolino, who had translated the whole file into English. After being given all kinds of assurances, Antonio even provided, upon request, a replica of his latest telettrofono, expecting he would start his experiments right away. Grant had promised he would "furnish the wires and means to put the telephone into use" and that he would notify them what day Meucci could start.

After about two months, since they had not been contacted as promised, Bertolino called upon Grant to verify that the necessary arrangements had been made. While the very first meeting had filled Antonio with hope, the second encounter ended with the prospect of postponing any experiment because of Grant's hectic schedule. As time went by, either Bertolino or Antonio himself regularly called at Grant's office at least every two weeks, to no avail.

Antonio became increasingly worried, if not anguished, for at least two reasons. First, he had left with Grant at ADT headquarters—though reluctantly—the whole portfolio consisting of his drawings, a description of his invention, and the method to put it into practice, detailed in a dozen double-sided handwritten pages; a copy of the caveat deposited and registered by the Patent Office; and an actual telephone set, specifically built for a possible test in situ. Second, and even more troubling, was the fact that Grant had been making himself consistently unavailable—either absent or, when actually in the office, stating through his secretary that there had been no time or opportunity to do anything yet.

Once again, Antonio started feeling betrayed. Filled with despair, he saw the world gradually crumbling around him and started to imagine the worst possible scenario. After so many years of hopes, dreams, and sacrifices, he was seriously at risk of having his main invention stolen.

With this in mind, Antonio managed to renew his caveat in 1872 and in 1873, thanks to the generosity of his faithful friend Angelo Bertolino, who by then realized he was somewhat responsible for not having protected Meucci. However, by the time of the next renewal on December 28, 1874, Antonio could not find anyone willing to give him even one more cent for his telephone. The caveat expired, leaving him with no legal claim to his invention.

He had come to the end of his rope. His friends preferred to help him and his wife with their living expenses instead—which was unfortunate because, in the words of the Patent Office Examiner in charge of receiving applications, "Had Mr. Meucci's caveat been renewed, no patent could have been issued to Bell."

After Grant had kept Antonio waiting for over two years, in 1874 Bertolino went to Grant's office once more on Antonio's behalf, demanding the immediate restitution of everything. Grant did not receive him, but he was curtly informed that there was no longer any interest in the project, that all the material had been misplaced or lost, and that if found, the office would see to it that it was returned.[30]

In fact, the prototype and the accompanying file Antonio had left with Grant in 1872 were never really lost. At various times Grant reportedly admitted that "he had, or thought he had, those papers somewhere" while saying to someone else that "the documents referred to had been returned." Only four years after Antonio had been dismissed, ADT was seen to be cultivating direct and continuing interest in the telephone business.[31]

Interestingly, Elisha Gray and Alexander Graham Bell, two independent inventors, were both researching multiple telegraphy during the spring of 1875 at Grant's ADT, which in the meantime had officially become part of the Western Union Group. Each man had been authorized to carry out field tests, assisted by the company's skilled engineers.

Gray, by then one of the most distinguished inventors in America, was forty years old and working out of his own firm, Western Electric, in Chicago. Unlike Antonio he had become very rich, successfully turning every patent he obtained into a commercial success. He had agreed to experiment at Western Union with the transmission of musical tones on their telegraph wires.

At the same time, after being appointed professor of vocal physiology at Boston University, Bell, the twenty-eight-year-old Scottish-born teacher of speech to the deaf, had also developed an interest in the transmission of harmonic vibrations. He obtained permission to experiment at Western Union's laboratories during the months of March, April, and May 1875.

Western Union had just announced that it was willing to pay up to $1 million to the inventor who could enable telegraph wires to carry multiple messages simultaneously, a way to ease the growing congestion. Telegraph companies were looking for methods to increase their message-carrying capabilities on the lines they had, rather than constructing new and expensive telegraph lines. Bell and Gray were in a neck-and-neck race to find a solution. The tangle of overhead telegraph wires in American cities was becoming overpowering and oppressive, almost blocking out the sky.[32]

Back in 1875, Gray was far better known than Bell. Twelve years his senior, he was recognized—at least in scientific circles—as one of the leading electrical researchers in the country after being granted about seventy groundbreaking patents around the world. He was born on a farm in Ohio, and at the age of twelve, had to quit school to go to work. Despite his lack of formal education, he was granted his first patent at the age of thirty-three for an improved telegraph relay. He gradually abandoned farming to fully dedicate himself to electrical experiments and building a career as a manufacturer and inventor.

His success led him to start a firm manufacturing high-quality telegraphic equipment. Western Union, a U.S. company holding a near-monopoly over the telegraph, bought a one-third interest in Gray's firm, making him a wealthy man. His company, under the new name of Western Electric, moved to Chicago and became the largest manufacturer of electric machinery in the country and the main supplier of equipment to Western Union.[33]

Bell, meanwhile, was a bright, accomplished, cultured, intensely driven young man of meager means and grand ambitions. He had inherited his interest in speech, sound, and the emerging field of acoustics from both his grandfather Alexander Bell and his father, Alexander Melville Bell, who was mentioned in George Bernard Shaw's *Pygmalion* as "the illustrious inventor of Visible Speech."[34] Visible Speech was an innovative attempt to systematically catalog all possible human vocal sounds by assigning each a written symbol that represented the placement of the tongue and lips as the particular sound was uttered. The system was used to aid the deaf in learning to speak.

Young Bell was schooled in elocution and high-society etiquette, both in Edinburgh, where he was born in 1847, and in London at his paternal grandfather's house. Rivaling Bell's vocal command and flair for public speaking was his natural ear for music. He was such a gifted musician that, by his early teens, he had been dreaming of becoming a concert pianist. Instead, he followed the career path laid out by his father, in a new era that marked the renaissance of acoustics, yielding dramatically new ways to explore the production, transmission, and perception of sounds. The telegraph had opened a world of instant communications, as well as vast areas of research that intrigued him.

His early job teaching music and elocution at a boarding school in Scotland gave him the opportunity to try to make his own mark in the new science of acoustics. He contacted leading authorities in phonetics and became a member of the Philological Society in London, which was concerned with the study of speech and language—a precursor of modern linguistics. Gradually the idea of sending sounds over the telegraph rooted itself firmly in his imagination, side by side with his deep concern for the deaf that would become his lifelong passion. Because of the untimely deaths of Bell's two brothers, both of tuberculosis, Bell's parents decided in 1870 to emigrate, worried about the health of their only surviving son.

Thanks to Bell's father's Visible Speech system, also vastly popular in North America, the family had no problem settling in Canada, in Brantford, Ontario. Soon after, at the age of twenty-four, Bell found his fortune in Boston, trading on his father's reputation. He was teaching deaf children and lecturing on vocal physiology and elocution at Boston University, which had just opened, while inventing in his spare time. Since he was eighteen, he had been focused on the idea that sound waves could induce sympathetic vibrations. While in Boston, he became aware that the telegraph industry was struggling to keep pace with the voluminous number of telegrams being sent. Therefore, his aim became to build a functioning multiple telegraph system to increase the number of messages that could be sent at the same time over a single telegraph wire.

Gray and Bell were rivals and were experimenting independently at the Western Union lab in New York on their electro-harmonic telegraph, trying to avoid each other, with a good dose of mutual suspicion. During this time, each inventor had separately been in touch with the in-house technicians,

George Prescott and Frank L. Pope, and actively counted on their collaboration. Apparently, at a certain point, the two head electricians had come into possession of Meucci's material. There is no way to determine how they received it. Without fully inspecting its content, they put it aside in the Western Union laboratory.

Frank Pope's brother Henry was the American District Telegraph Company superintendent. One day, according to Frank's deposition ten years later, during the 1885 Bell-Meucci trial, Henry had brought home "two instruments." Was this the set that Meucci specifically built at Grant's request? Or maybe a prototype Ester had sold when her husband was bedridden? It remains a mystery. What was revealed during the trial is that Western Union president William Orton, while encouraging further experiments from young inventor Thomas A. Edison, had ordered the Pope brothers to evaluate the feasibility of this invention "in the frame of a 360-degree survey of the telephone." Frank and Henry Pope spoke on that very phone for two or three hours, back and forth, on a wire they installed between their houses. This set could not have belonged to A. G. Bell, since he did not have a telephone ready for demonstration until months after his patent was granted in 1876.[35]

Equally mysterious are the relationships of Pope and his colleague Prescott with Bell and Gray. There is no proof that they allowed either inventor to consult Meucci's papers or to test his perfectly working prototype, left at their offices. However, the sheer number of coincidences in this case makes it impossible not to grow suspicious.

What is sure in this web of mysteries, lies, and embroilments is that neither Gray nor Bell initially intended to work on the electrical transmission of speech. After working for a few months at the Western Union laboratory on their harmonic telegraph, trying to transmit musical sound and even tunes, each shifted his field of research and started to experiment with the electric transmission of the human voice. How could this have happened?

As a matter of fact, when the telephone came into being, the prevailing mentality was total skepticism. Little more than a hundred years ago the transmission of voice at a distance, which we nonchalantly take for granted today, was considered a preposterous hypothesis. A few select inventors knew that it was indeed possible. In 1854 Johann Philipp Reis was the first to use the word "telephone" and to believe in the feasibility of the "electrical transmission of the spoken word," but, being a government employee afraid to lose his job, let the whole thing go, allowing bureaucratic myopia to delay the whole process by more than two decades. But Meucci, though he did not use the word "telephone," had invented his telettrofono five years earlier, in 1849 in Havana, and had then made it operational in Staten Island.

Since both Gray and Bell were experimenting at Western Union, they may well have had a chance to consult at their leisure Meucci's precious papers, drawings, and prototype, which offered unique new perspectives, soon grasping the concrete possibility to make a major revolution in the world of communications.[36]

Each of them, in fact, presented a patent and a caveat, respectively, on the same day, February 14, 1876, a little over a year after Meucci's caveat had expired. Bell's application for a patent on "Improvements in Telegraphy" was rather imprecise and full of inconsistencies. It did not mention the word "telephone" or promise the electrical transmission of speech, but only of "vocal or other sounds." Gray's caveat for an "Electric Speaking Telephone," however, was impeccably clear and focused. One is inevitably struck by the similarities of both Bell's and Gray's descriptions to the apparatus devised by Meucci.

Antonio, after almost thirty years of relentless and dedicated research, had given away his whole invention in good faith. He was convinced that he would be the protagonist of in-house experiments at Western Union, which would confirm his own theories. Instead, he ended up becoming the provider of an immense trove of information for someone else's gratification and financial reward. He was blatantly deceived, along with his notary public, Angelo Bertolino, and ultimately robbed.

Perhaps Grant was not lying when he stated that he had entirely forgotten that Meucci's detailed proposal had been handed over to his staff. However, it's very unlikely that George Prescott and Frank L. Pope forgot they had allowed Gray and Bell to consult Meucci's papers and test his telephone. Prescott's main task at Western Union was to screen the flood of inventions submitted to the attention of the company. All telegraphic innovations of the time could be put into practical, operational shape only with his direct approval. Pope, Thomas Edison's former mentor and partner, specialized in electrical patents and was a real authority in evaluating new inventions. His memory became oddly blurred, especially after accepting a position at the Bell Telephone Company.

The whole attitude of Grant's consortium of unscrupulous and equivocal people showed total disrespect toward the Italian inventor and his legal representative. The ordeal of pointless waiting that Meucci was forced to endure was definitely rooted in opportunism, if not outright prejudice against Italian immigrants, even those who had fully integrated into the establishment through competence and with the required legal credentials.

On the other hand, it is a fact that Antonio, after filing his first caveat and having it registered at the end of December 1871, felt so protected by the law—unlike he had before that date—that he made no secret of his invention, not even with some reporters. An accurate description of it was published in a New York newspaper, possibly the *Tribune*. After that, he was asked by William Carroll, a diver who was perfectly aware of his invention, if he could adapt it into a marine telephone, allowing underwater communication.

New York at that time was one tenth of the size it is today, so Antonio was quite well known both in the city center and in Staten Island, where his Clifton cottage had become a point of reference for Italian Americans. He felt free to describe the telephone he invented to his friends and allowed a number of people to test it in his own home, where he used it daily. His open and spontaneous behavior, especially at the Italian Consulate, was perhaps due to the fact that he felt linguistically and emotionally at ease with his compatriots. All of them were often looked down upon as second-rate citizens, despite their legitimate aspirations to dignified lives in their adoptive country.

Antonio strove all his life to fight all kinds of obstacles, including the ingrained skepticism of his contemporaries toward his main invention. A. G. Bell, on the other hand, was truly born under a lucky star. While teaching Visible Speech at the Boston Horace Mann School for the Deaf, he became close to Thomas Sanders, a rich businessman who made a fortune in animal breeding. He offered Bell free room and board at his stately home in Salem, Massachusetts, in exchange for his tutoring services. The eldest of Sanders' eight children, George, was born deaf, and Bell agreed to give him private classes in speech using his innovative method for the deaf rather than the limiting standard sign language.

Soon after, in 1873, on Sanders' suggestion, Graham also started teaching the sweet and vivacious sixteen-year-old Mabel, the deaf daughter of Gardiner Greene Hubbard. The son of a Supreme Court justice, Hubbard was a powerful patent lawyer with offices in Boston and Washington, D.C. Thanks to his experience as a professional lobbyist, he was appointed head of a special commission on mail transportation by the president of the United States, Ulysses S. Grant, and later reconfirmed by President Rutherford B. Hayes.

Bell and Hubbard first met in 1874 at the Hubbard residence in Cambridge's wealthiest neighborhood. They soon realized they had a lot in common. They shared the same view about the education of the deaf, in particular of Mabel, who had been left permanently deaf after a severe case of scarlet fever at the age of three. Bell knew of Hubbard's interest in telegraphy long before visiting his household because of the notorious Hubbard Bill.

Convinced that Western Union was overcharging its customers and therefore hurting the U.S. economy, Hubbard had introduced a proposal to Congress to dismantle the existing monopoly and nationalize the telegraph industry. His idea was to create a private corporation, a consortium led by himself and his business associates in the service of the federal government. The bill sparked quite some controversy and upheaval, but ultimately failed to pass, due in part to William Orton, president of Western Union and Hubbard's arch enemy, who bribed many members of Congress by offering a precious card that granted unlimited free telegrams. The political climate at that time was pervaded by intimidation and corruption, culminating in 1876 with the controversial election of Republican Rutherford B. Hayes as the nineteenth U.S. president.

As soon as Bell told his host that he believed he could build a telegraph system capable of sending multiple messages over a single wire simultaneously, Hubbard's entrepreneurial imagination was set alight. He saw an opportunity that might allow him to rival Western Union by launching a new telecommunication network. They soon had some lengthy discussion about Bell's experiments and his need for funding. He had already produced the first crude drawings in the pursuit of a "harmonic or acoustic telegraph" in 1872, which would enable multiplexing telegraph messages simultaneously over a single wire by using different frequencies for each message. By late 1874 Hubbard became more interested in Bell's elaborate experiments and made an unexpected and daring search at the Patent Office in Washington, D.C., to verify if there were other pending patents in the area of the transmission of sound.[37]

It's very possible that thanks to his lawyers, Anthony Pollok and Marcellus Bailey, partners at one of the capital's leading law firms, Hubbard came across Meucci's caveat, which was filed under the category of Sound Telegraph and was about to expire. Oddly enough, just after Hubbard visited the Patent Office, Bell suddenly considered for the first time, in November 1874, the idea of constructing a device to actually transmit speech. Moreover, in a letter to his parents, he declared:

> *I am tonight a happy man. Success seems to meet me on every hand . . . After my last interview* [with Hubbard] *he had gone down to Washington, and searched the Patent Office to find whether my idea had been taken up by anyone else—and he now offers to provide me with funds for the purpose of experimenting if we go into the scheme as partners.*

Bell was also finally allowed to have an assistant who could help him build his electrical prototypes: Thomas Watson was an enterprising, earnest young man, whose little formal education had not prevented him from successfully working at the iconic Charles Williams machine shop on Court Street in Boston. This was a thrilling hub for visionary inventors, including young

Thomas Edison, who was by then the same age as Bell, and was attempting to build all sorts of devices exploring the still little-understood potential of electricity.

Hubbard's steady interest in telegraphy and complete trust in the unique method of his young daughter's private teacher encouraged Bell to become a frequent visitor to the Hubbard home and a regular guest at the family's exquisite midday Sunday dinners. Over time, he became increasingly infatuated with his student Mabel, eleven years his junior, and more indebted, both financially and emotionally, to his prospective father-in-law.

Hubbard was perfectly aware that speed was essential if Bell were to beat out his competitors with his new telegraphic invention. He knew through his Washington lawyers that Elisha Gray, affiliated with Western Union, intended to file a caveat and therefore tried to urge Bell to give up his teaching and devote himself full-time to research. Upset by this proposal, Bell redoubled his efforts without relinquishing his vocation as a teacher.

On February 27, 1875, a significant and consequential verbal agreement with the parents of his two private students, Sanders and Hubbard, took place. This came to be known as the Bell Patent Association. Bell's inventions would be underwritten by his two partners, who would provide the legal fees and financial resources necessary to carry on his experiments in exchange for a share of any potential profit. Hubbard was to be the trustee or president, and Sanders the treasurer. From the very beginning Hubbard insisted that a patent rather than a caveat be filed as soon as possible to better undermine Gray's claim.[38]

Despite the agreement to jointly provide $50,000 for the start of the telegraph and telephone business, it was Sanders who paid all the expenses since Hubbard, though very wealthy, had no available cash as he had already invested in major civic projects, such as the first street railway line outside the New York City area and the Cambridge Gas Light Company. Given his reputation as an enlightened public man and a zealous entrepreneur, Hubbard was in a position to provide priceless connections at the Patent Office and at the highest levels of government. A year later he decided to move permanently to Washington, D.C., to lobby Congress on railroad mail transportation.

He had made sure that Bell, on his first trip to Washington on February 16 to March 2, 1875, would meet with Hubbard's legal representatives, Pollok and Bailey, "the most eminent men connected with the Patent Office." It was at their office that Bell first met Zenas F. Wilber, the examiner at the U.S. Patent Office, only half a block away.

In today's practice, a patent examiner is not allowed to pay visits to a patent lawyer's office or discuss matters concerning a patent application before it is filed. But back then, things were different. And Wilber had a good reason for making a visit. He was chronically in need of money to satisfy his drinking habits, and apparently Marcellus Bailey, with whom he had served in the Civil War, was always ready to grant him a loan without ever asking to be repaid. Bailey had more influence over Wilber than any other attorney in Washington.

On that very occasion, Pollok and Bailey advised their new client Bell to split his patent application on multiple telegraphy into three distinct applications. They'd had the occasion to privately examine, at the Patent Office, the pending application just filed on the same subject by Elisha Gray. This strategy would guarantee that at least one application would pass right away. In fact, the most important one, relating to Bell's multiple telegraph and filed on March 6, 1875, was granted in record time within a month, thanks to the direct intervention of Hubbard's team of lawyers.

During this same trip to Washington, Bell's interest in the electrical transmission of speech grew stronger than ever. After Hubbard's recent inquiry at the Patent Office, Bell paid two visits to physicist and inventor Joseph Henry at the Smithsonian Institution to discuss the subject at length. While there, he took the opportunity to see the Johann Philipp Reis telephone, both the transmitter and the receiver.

By this time Bell had conceived a so-called harp apparatus for his harmonic telegraph, which he thought could be altered slightly to transmit speech. He described it with great confidence in a letter to his parents after returning home from Washington. He made a couple of trials on his own, but with no success. After a frustrating summer, distracted and torn by his love for Mabel, in October he started working productively again, thanks to his assistant Thomas Watson, on a new patent that would later become "the most valuable patent in the history of mankind." He was in no hurry because he intended to file the patent in England and only a caveat in the United States. Born in Scotland and not yet an American citizen, Bell strongly felt he belonged to the United Kingdom. He had contacted George Brown, a father of the Canadian Confederation, to help him secure the patent in Britain and was willing to wait for his reply.

On Mabel's eighteenth birthday, November 25, 1875, the two finally became formally engaged with her family's approval. Bell was deeply in love but anxiously aware of the huge gap between Mabel's lavish lifestyle and his own meager means as a mere professor and teacher. As his faithful assistant Watson wrote later on, in his autobiography:

> *Professor Bell had a special trouble all to himself . . . he had fallen in love, wanted to get married and didn't have money enough. The important question for young Aleck was, where could he get the money that would enable him to marry his beloved?*[39]

How could he improve his financial position? In Bell's correspondence to his parents, he confided:

> *It is a neck-and-neck race between Mr. Gray and myself who shall complete our apparatus first. He has the advantage over me being a practical electrician—but I have reason to believe that I am better acquainted with the phenomena of sound than he is—so that I have an advantage there . . . I feel I shall be seriously ill should I fail in this now I am so thoroughly wrought up.*

By then Elisha Gray and A. G. Bell had very good reasons to suspect and anticipate each other, as they were working on the same invention. So Hubbard instructed the Pollok & Bailey firm to go for a hasty filing, and rushed on "without Bell's knowledge or consent."

It was on this crucial decision that Bell's life would turn from rags to riches. He didn't know that his application had been filed on February 14, 1876, until after the fact. The exact events that occurred on that day and shortly after will never be known for certain. What we do know is that on the same day, Gray filed a caveat application for a speaking telephone.

Gray, unlike Bell, had already proven to be one of America's most successful inventors of electrical devices. He was applauded by *The New York Times* with a flattering article entitled "Music by Telegraph." Hubbard did not want to risk Gray acquiring the significant advantage of a granted caveat.

How coincidental that Meucci's caveat had so recently expired! Given Pollok & Bailey's strong connection with the Patent Office, it would have been so easy for them or Bell to inspect that document. In spite of the 1870 Patent Act, which clearly specified that "preservation in secrecy was prescribed indefinitely and unconditionally," Bell or anyone else could transcribe all or part of Meucci's specification at their leisure.

At that time, American patent laws were structured in a way that allowed perjury, bribery, misdating, and other trickery. Hubbard's strategy turned into a true gambit, a very well-focused and practical conspiracy that exploited a safe, reliable contact at the Patent Office, Zenas Wilber. Moreover, it would have been very hard to prove that there had been a breach of secrecy, even though government corruption was a given fact at the time.[40]

What we do know is that attorney Anthony Pollok, from the firm Pollok & Bailey, filed an application for Bell's Improvements in Telegraphy patent, while Gray filed an application for a Speaking Telephone caveat on the same day. The distinction between them is stark.

Gray's caveat contains a single claim, supported by the accurate description of the liquid transmitter with specific drawings:

> *I claim as my invention the art of transmitting vocal sounds or conversations telegraphically through an electric circuit.*

Bell's patent, by contrast, includes five claims. The first four items claim the invention of a multiple-messaging telegraph using "undulations in a continuous voltaic circuit." The fifth and most famous claim is incorporated in the final sentence of the patent, as the only mention of the transmission of vocal sounds by telegraph:

> *The method of, and apparatus for, transmitting vocal or other sounds telegraphically, as herein described, by causing electrical undulations, similar in form to the vibrations of the air accompanying the said vocal or other sounds, substantially as set forth.*

Bell's patent includes only a few lines, handwritten into the left-hand margin, hinting at the possibility of a liquid transmitter as to a possible further invention.

While Gray's caveat is clear and direct, describing an instrument to transmit speech with a precise design that used a water transmitter, Bell's patent application mainly describes a telegraph system capable of sending multiple messages at the same time. It makes no specific mention of transmitting speech and no prototype was enclosed. Why didn't Bell emphasize the possibility of transmitting speech in the body of such a major patent, rather than making it sound like an afterthought?

While the application and fee for both inventions were duly recorded by the clerk, the time of the entry was not reported. Had the patent clerk noted the hour when each of these applications had been filed, years of litigation would have been avoided.

What we know for certain is that Bell's application went directly to Examiner Zenas Wilber's office, while Gray's caveat remained in the in-basket until the end of the day, as was customary. So whoever submitted the patent application on Bell's behalf had bypassed normal procedure, insisting it be immediately hand-delivered to the office of the patent examiner.

Wilber was not suited for this sensitive and delicate job. He was a known alcoholic and had been arrested more than once for drinking, but as the cousin and ward of Rutherford B. Hayes, soon to be elected the American president in 1876, he managed to keep his position at the Patent Office.

It was certainly very strange and suspicious that Bell's patent was passed in less than three weeks, something of a record, and without any objection when every claim in it was amenable to criticism under office rules of practice. As of February 14, the

Examiner's Office, room 118, became an unusually busy destination.

When Wilber received the two applications, he saw right away some overlapping areas of concern and therefore placed them in temporary suspension, due to likely interference. He then notified Gray as well as Bell and his lawyers. Pollok & Bailey immediately protested, maintaining that their client's application was filed before Gray's and therefore no declaration of interference should be allowed. In his defense, Wilber explained the reasons that led him to declare the interference, inadvertently revealing the contents of Gray's caveat. Upon learning this information, Bailey telegraphed his client and acutely reminded Bell that a fortune was at stake. Following his lawyers' advice, Bell rushed to Washington to call on Wilber to clarify the nature of the interfering claims.

Maybe Bell wasn't too well versed in the law or the ethics of patent procedure and didn't realize he was asking for a breach of official confidence that could give him a clear advantage over his competitor. Wilber was a deaf-mute who knew of Bell's work with the deaf and had previously met him informally at Pollok & Bailey's office. The patent examiner, given his ties with Bailey, shared key information of great relevance.

The commissioner of patents, Robert H. Duell, who had consented to the lawyers' convincing objection, asked Wilber to withdraw the suspension. So Wilber gave precedence to Bell's application without making any further inquiries, though the records in the office did not show which was filed first on the same day. The official version ended up stating that on

February 14, 1876, Bell's patent was the fifth received and Gray's caveat was the thirty-ninth.

In Gray's view, Wilber stood as the pivotal insider with the ability to facilitate Bell's plagiarism by letting him know the technical content of Gray's caveat, and even allowing him to amend his patent application.

At least two questions come to mind at this point. Why wasn't Bell required to submit a working model for his invention? And why wasn't Bell's apparent plagiarism from Gray's and Meucci's caveats uncovered and challenged in the following years of bitter court battles?[41]

A few weeks before the patent application was filed, Bell complained to Mabel in a letter about all the work he had to do. Three identical copies were needed: one for the U.S. Patent Office, one for himself, and the last one for George Brown for the patent application he meant to file in England. It turned out that Brown's copy of Bell's patent application did not include any notes written in the margins. Most likely, such notes were also absent from the copy that Bell swore to before a notary public on January 20, and from the version that Pollok & Bailey filed with the Patent Office on February 14.

At that time, the "secret archives," as they were called, were not at all secret. Rooms were not locked, doors were always open, and caveats and applications were kept in pigeonholes and drawers. All of it was accessible to anyone entering the building, who could even take files from the office and sometimes keep them overnight. Examiners were offered money to secure favorable outcomes.

This seems to be the case with Bell, who, after several prolonged visits with Wilber, presented the patent examiner with a $100 bill, a very substantial amount at that time, to show his appreciation just a few days before his patent was granted. This patent for improvements in telegraphy soon became known, oddly enough, as the patent for the invention of the telephone.

From that time on, the story of the invention of the telephone became an increasingly tangled web of patents—"a story of fraud, collusion, perjury, corruption, bribery, and what would now be called industrial espionage."[42]

Bell's attorneys had an underground railroad operating between their office and Examiner Wilber's room at the Patent Office. They felt authorized to withdraw Bell's application, change it (even including Gray's variable resistance method) over Bell's signature, and then return it to the files as though it were the original. All of this puts into question the professional integrity and moral character of eminent Washington attorneys, as well as of the examiner who allowed all of this to happen through ignorance, misinterpretation, and corruption, often reported to the U.S. Congress but conveniently ignored.[43]

While Meucci had spent almost thirty years building an efficiently working model with his own hands, which was used every single day by his immediate family and friends, Bell took a rather different approach. Thanks to Gardiner Hubbard's timely investigations and manipulative nature, plus the tentacular influence of his Washington lawyers, Bell found himself the recipient of a registered patent on March 7, 1876, for a telephone that did not yet exist. And thanks to his attorney Bailey's influence, Bell was then able to amend his application and obtain the privilege of creating an invention at a later date that he had not yet fully thought through. In fact, it was three days after he had received the patent that Bell pronounced the famous phrase "Mr. Watson, come here!"—achieving success for the first time in transmitting the human word over an electric wire, with the fundamental help of his assistant.

On March 10, as specified by Thomas Watson in his detailed diary and on a formal plaque outside Bell's workshop at 5 Exeter Place in Boston, "Alexander Graham Bell transmitted to Thomas Augustus Watson the first complete and intelligible sentence by telephone." This is what most American history books report, possibly oblivious that in Havana, on New Years's Eve 1849, to the astonishment of all his guests, Antonio Meucci announced and demonstrated that his telephone was truly working.

Bell's successful transmitter, however, did not bear any resemblance to the one described in his approximate and confusing patent application, but was strikingly similar to the one in Gray's caveat. When Bell decided to go public and give his first lectures at the American Academy of Arts and Sciences on May 22, and at the Massachusetts Institute of Technology on May 25 of the same year, he limited himself to showing the device rather than demonstrating its use, cleverly avoiding having to recount the story of his first success in transmitting intelligible speech.[44]

As a matter of fact, in early 1876, Bell still had no clear insight into the real requirements for the transmission of speech. After his "telephone" patent was issued, his prototypes continued to be inadequate for the next seven months. Despite his efforts, he could not get his "invention" to work satisfactorily—or work at all—because, unlike the vowels, the consonants were barely discernible.

In a moment of spontaneous humbleness, Elisha Gray admitted that his own instrument was not working too well and openly wrote to his rival:

> *I do not claim, even the credit of inventing it* [the telephone], *as I do not believe a mere description of an idea that has never been reduced to practice—in the strict sense of that phrase—should be dignified with the name invention.*

In June 1876, Bell was given a unique opportunity: He was invited, along with the official exhibitors, to the grand Centennial Exposition in Philadelphia—the first World's Fair held in the U.S.—celebrating the one hundredth anniversary of American independence. This opportunity, too, was meticulously orchestrated by Hubbard, who, being a respected longtime member of the Board of Education of Massachusetts and of the Exposition in Philadelphia, had clear ideas on how to promote newly patented inventions as profitable business ventures.

Even though all the display booths at the trade fair were already reserved, Hubbard was able to secure an additional booth in the education section, within the specific context of the education of the deaf, with a distinct display of Visible Speech books and charts. He also made sure Bell could meet the three judges of the electrical department of the Exposition beforehand,

and finally show his telephone and multiple telegraph to some of Hubbard's influential acquaintances. Bell had been reluctant to participate at this late date, but with the insistence of Hubbard's daughter, Mabel, he agreed not to miss this chance.[45]

Though Bell and his harmonic telegraph and telephone were peripherally displayed, the emperor of Brazil, Pedro II, came by during the judging session of the electrical devices on Sunday, June 25, and recognized the young professor he had met two weeks earlier in Boston at the School for the Deaf. Accompanied by Sir William Thomson, a famous British physicist, the emperor tried the instrument and was transfixed because it talked. Though it only pronounced "to be or not to be," the six monosyllables impressed the monarch, and Bell became famous overnight. He had cleverly specified to the audience, including Elisha Gray—who had been prominently featured by Western Union—that the transmission of the human voice was still "an invention in embryo." Pedro II was the first person to buy stock from the Bell Telephone Company, established a year later, and had a phone installed in his summer retreat forty miles from Rio de Janeiro.

When Meucci found out from the press that the great invention of his life bore someone else's name, he cried like a baby, as his friend Domenico Mariani recalled. It was as if his child had been kidnapped. In his own words, this was his first reaction: "I saw that the patent for the invention of the telephone had been granted to a certain A. G. Bell. I immediately recognized that my invention, the fruit of many years of labor, had been stolen."

Chapter Five

ANTONIO FIGHTS BACK
(1876–1889)

After he learned that Bell had been granted the patent for the telephone, Antonio was incredulous and speechless—for a while. But soon his inertia turned into fierce determination to defend his priority in the telephone invention.

He immediately wrote to his lawyer, Thomas Stetson, urging him to inform the Patent Office in Washington that the invention for which he was granted a caveat in 1871 had been stolen from him. He also asked Angelo Bertolino, equally deeply offended by the turn of events, to write to the newspapers that Meucci's legitimate rights over his invention of the telephone had been blatantly robbed.

Antonio decided to invest all of his energy to defend the ownership of his telettrofono. In an attempt to see his rights recognized and obtain official support from Italy, he informed everyone he could think of, including Garibaldi in his retreat on the island of Caprera.

The Meuccis' house during that time saw a constant coming and going of all kinds of people—candid idealists who hoped that truth would prevail, curious neighbors, and the usual profiteers. Bertolino introduced him to the apparently well-intentioned Colonel William Bennett, who prepared a promising contract. It obliged Antonio to reproduce all the instruments he had designed in the last twenty-five years, unfortunately sold by his wife, in exchange for an adequate advance and the promise of a weekly refund for his expenses. Antonio managed to rebuild all the instruments, thanks to the generosity of Ashael K. Eaton, a very fine scientist and professor of physics, who invited him to his laboratory in Brooklyn, where he was given all the material he needed. In the end, these efforts proved useless because Colonel Bennett disappeared.

During this period, to survive, Antonio patented a lactometer for testing milk (September 28, 1875: U.S. Patent #168273) and a hygrometer for measuring humidity (October 10, 1876: U.S. Patent #183062), assisted by a friend, Giuseppe Tagliabue, who paid for the expenses. The income from these two inventions was so little that he decided to produce a new kind of ornamental candle for Christmas, which sold well. When he was finally getting back on his feet, another professional swindler came along: Charles Raudwitz, who placed a large order and disappeared without paying.[46]

While trying to assert his ownership of the telephone, Antonio continued to explore his wide range of interests and, above all, to invent. He devised a formula for a super effective glue and a system of noise prevention for elevated railways.

What was unique about Antonio was his modus operandi. He would conceive an idea, test it, and construct the final product all by himself. Bell, instead, had always counted on a number of technicians, assistants, and consultants, in particular Thomas Watson. He also had more than one lawyer prepare his patent applications.

Despite successfully obtaining the patent, Bell's instrument did not yield the expected results. While testing it in Scotland, Sir William Thomson found that the vowel qualities could be articulated, but no consonant sound could be properly heard.

While visiting his parents in Canada from August 10 to 12, 1876, Bell conducted many experiments. The quality of speech transmission, however, did not substantially improve. Since consonants were still mostly unrecognizable, speech quality remained marginal.

Satisfactory results, as finally confirmed by Sir William Thomson, were ultimately achieved by Bell and Watson in the first two-way reciprocal conversation on October 9, 1876, over a two-mile line between Boston and Cambridge. Their success was mainly due to the use of membrane diaphragms, which were to be soon followed by metallic ones, as in the "box telephone" that Bell constructed between October 19 and 20, 1876, and fully described in his second patent, filed on January 15, 1877. This patent was granted with extraordinary rapidity, in just fifteen days.

Some distinctive improvements were clearly and diligently specified in the application:

1. a plate of iron or steel is substituted for the membrane-with-glued-armature diaphragm, in both the transmitting and receiving instrument
2. positioning the coil at the end of the magnet nearest the diaphragm
3. an adjusting screw is added, whereby the air gap may be finely adjusted to a minimum, but non-zero value
4. the need of a call bell is recommended, though no construction specifications are specified
5. the use of a permanently magnetized core, allowing to eliminate the battery is envisaged.

In his Memorandum Book, Antonio wrote an important note dated May 20, 1862: "At the centre of the wire, a strong magnetic iron protected by a bobbin do not need any battery at all and is a good conductor of the sound." This means Antonio had experimented with permanently magnetized cores since 1862, fifteen years before A. G. Bell.

The substantial improvements of Bell's second patent in 1877 did not prevent the inventor, two days before filing the application, from still publicly complaining before the Philosophical Society in Washington about the persistent "difficulty with such consonants as *p*, *t*, and *k*."

There would be still a long time before Bell's results could challenge those achieved by Antonio. In fact, to see a circular iron diaphragm firmly clamped along its rim—though not as well as in Antonio's soap-box telephone of 1864—we'd have to wait until May 1877, when Bell exhibited it before the Society of Arts in Boston.

The same is true for other recommendations suggested by Meucci in his 1871 caveat. As far as a first call signaling, first envisioned in 1858, we'd wait until 1878. To see a telephone line coping with the skin effect, which Antonio investigated from 1862 to 1870, we'd wait until the 1880s. Finally, to see the first anti-sidetone circuits and the inductive loading that Antonio had devised in 1858 and 1862, respectively, we'd wait until the next century, when these features were introduced by Michael I. Pupin and George A. Campbell, not by Bell.[47]

After Bell had obtained the three major patents, Hubbard thought it was the right time to solidify his control over his future son-in-law's inventions. He decided, as a true deus ex machina, what to do and when on Bell's behalf to gradually secure exclusive rights to the telephone—opening the way to what would become one of the largest and most lucrative monopolies the world has ever known. Hubbard traveled extensively throughout the U.S. on behalf of the government and brought a prototype of Bell's "phone" everywhere he went, giving demonstrations left and right.

Thanks to a group of capitalists who were confident enough to believe that the telephone could be rapidly improved and soon be a success on the market, on July 9, 1877, the Bell Telephone Company was formed, with Hubbard as trustee or president, as anticipated, and Thomas Sanders as treasurer. Five thousand shares of stock were issued. They were divided as follows: Gardiner Hubbard: 1,397 shares; Gertrude Hubbard: 100; Alexander Graham Bell: 1,497; Thomas Sanders: 1,497; Thomas Watson: 499; and Charles Hubbard: 10.

When Bell married Mabel Hubbard two days later, on July 11, he gave all but ten shares of his stock to Mabel as a wedding present, which would allow the couple to live in comfortable splendor for the rest of their lives. Hubbard was given the power of attorney by his wife and daughter, and thus controlled 60 percent of the company's stock, and Bell only 20 percent.[48]

After their intimate wedding at the Hubbard residence, Bell and Mabel left for Europe for a fifteen-month trip. While in London, Bell relished the chance to demonstrate the telephone personally to Queen Victoria and accept countless honors. In Italy his patent was registered on November 27, 1877, and a month later the first local call took place in Milan. On February 28, 1878, the first long-distance call between the Quirinale Palace in Rome and the Post-Telegraph Office of Tivoli was successfully celebrated in the presence of the royal family. Seventeen years had passed since Meucci had first tried to involve the Italian authorities in his innovative project, through his spokesman, Enrico Bandelari. Unfortunately, when Antonio had tried, nobody was prepared to listen to a proposal so radically ahead of its time.

Once back in the U.S. in 1878, Bell took no further interest in the Bell Telephone Company. On some level, he must have been deeply embarrassed, if not ashamed, about the circumstances that led to his patent. Why, for instance, did he give his stock in the Bell Telephone Company to Mabel? Aside from testifying on the Bell Telephone Company's behalf at Hubbard's insistence, why did he refuse any personal or financial involvement in the newly formed company that bore his name? He made sure there would be no connection with himself whatsoever, except as a stockholder. Maybe it was an elegant way to let his father-in-law know that the Bell Telephone Company was Hubbard's own creation.

Bell had decided to build a lavish retreat on Canada's Cape Breton Island—to get "far away" from the looming controversy over his disputed role in the invention of the phone—when his life was shipwrecked.

The first major attack on the Bell Telephone Company in its intent to monopolize the promising telephone business came in 1878 from the American Speaking Telephone Company, headed by Peter Dowd. The company was an offshoot of Western Union that had begun offering telephone service based on patents licensed from Elisha Gray, Thomas Edison, and several others. The president of Western Union was indeed interested in the telephone, but not in Bell's invention, since he was aware it was compromised. The case against Bell maintained, first, that Gray, Edison, and others had priority of invention; second, that Bell's instrument as described in Patent No. 174,465 could not transmit articulate speech; and finally, that Bell's invention had been anticipated by a list of prior publications in the public domain, and therefore was not patentable.

In order to protect its interests, the Bell Telephone Company, now registered as the National Bell Telephone Company, tried to shut down the American Speaking Telephone Company by suing

Dowd, and by proxy the Western Union Telegraph Company, for patent infringement.

The case was the first to bring the telephone-related claims of Bell and Gray into conflict at court. It was therefore crucial, according to the rules of the day, that A. G. Bell make a pretrial deposition. In spite of Hubbard's insistence, Bell did everything to avoid testifying under oath about his role in the development of the telephone. He informed Hubbard that he and Mabel and their new baby daughter, Elsie, would be coming back from Europe and would stop in Canada, bypassing Boston altogether.

Hubbard, "the power behind the throne," was determined to protect the monopoly he had started to establish and to win at all costs against Western Union. In an effort to save the financial power of his institution, he wrote to Bell's father for help, dispatched Watson to reach Bell wherever he was, and implored Mabel to convince her husband to return to Boston. During the more than 600 legal actions that followed, whenever Bell was asked to prove that he was the inventor, he would never give a straight yes-or-no answer or swear under oath that he had invented the telephone.

Too often, historical records seem to indicate that the Bell-Western Union conflict ended with a judicial decision, which is not true. It ended with an agreement and a settlement. To everybody's surprise the agreement was signed on November 10, 1879, and the settlement was confirmed by the court. When Meucci discovered that Western Union agreed to transfer its telephone patents and withdraw from the business by selling its 56,000 telephones to the National Bell Telephone Company, he became aware that Bell was backed by a powerful and efficient organization.

Western Union was bought at a very high price. They had to recognize Bell as the inventor of the telephone, and the National Bell Company committed to pay Western Union and Elisha Gray 20 percent of all profits from telephone rentals or royalties for seventeen years.

So why did Western Union leave the field free for the National Bell Telephone Company to become an empire, considering that it had already installed so many phones in fifty-five American cities and was rapidly and profitably expanding? This is a question with no easy answer. Meucci, however, understood right away that this was an opportunistic agreement of reciprocal economic convenience. To put it simply, Western Union already had a profitable telegraph monopoly. A fight with Bell would have been costly and lengthy. All things considered, since Bell agreed to keep out of the telegraph business, Western Union gave up the plan to extend its monopoly to telephones.

Bell's priority as the inventor of the phone was not confirmed by the court, but thanks to a shrewd financial arrangement, it became automatically assumed. The terms of the November 1879 settlement dictated that Western Union recognize A. G. Bell as the actual and the first inventor of the telephone.

This occurred despite testimony from Frank Pope and a publication from George Prescott. Western Union's chief electrician, Pope testified in favor of Elisha Gray (but subsequently attributed the invention exclusively to Bell after he left Western Union and started working for the Bell Telephone Company).

Similarly, Prescott, the electrical researcher who served in 1878 as Western Union's chief engineer, with a reputation for fairness and independence, had published a book entitled *The Speaking Telephone, Talking Phonograph, and Other Novelties*, in which he credited Gray with the invention (but in the 1884 edition, thoroughly rewritten, he praised Bell's "original discovery"). These examples confirm how certain kinds of erroneous historical myths are generated.

The immediate result was the dramatic rise of the National Bell Telephone Company in the stock market . The price of Bell stock rose overnight from $110 to $995 a share. From the very beginning the company chose to pay exorbitant amounts of money in legal battles, but nonetheless had turned into a powerful financial institution. Hubbard managed the business like a dictator, much to the irritation of his partner, Thomas Sanders, gradually revealing himself for what he was: a true robber baron, in the same league as John D. Rockefeller, Andrew Carnegie, J. P. Morgan, and Cornelius Vanderbilt. By the 1880s, corruption and bribery were necessary ingredients to succeed in the formation of monopolies, until the Sherman Antitrust Act of 1890, which was nevertheless conveniently ignored.

The National Bell Telephone Company aggressively succeeded in destroying opposition in the telephone business using devious monopolistic tactics. Though Hubbard remained the controlling force within the corporate structure, he surrendered his absolute voting majority and hired Theodore Vail from the Postal Service to be the general manager in charge of operations.[49] Vail's relentless Yankee corporate executives hastily swept most of the new telephone companies into a nationwide monopoly. These individuals would reportedly stop at nothing in a scheming campaign of greed under the appearance of legal democratic service to the people. At least one poignant specific example, among many others, of the Bell Company's "robber baron" strategies is worth quoting from Paul Latzke's enlightening publication, *A Fight with an Octopus:*

> *As soon as a decision in this case came down, the Bell people proceeded to stamp the life out of . . . the competitors. In a dozen states, injunctions were secured and infringement damages promptly assessed. Under these judgments telephone plant after telephone plant was ripped out, and the apparatus was piled up in the most conspicuous place that could be found and burnt in a public bonfire as an "objective lesson."*[50]

A. G. Bell was gradually acquiring great prestige and his invention was recognized as the first and only official telephone.

When Meucci fully grasped the implications of the 1879 Bell-Western Union settlement, he knew he had to fight. He couldn't watch the glorification of a man who had stolen his invention. In spite of the sense of loss and grief he experienced after the recent death of his friend Lorenzo Salvi and, even more so, of his heroic mentor Giuseppe Garibaldi, Antonio found the stamina to react and stand for his own rights as inventor.

With the help of his notary public Angelo Bertolino and his lawyer Michele Lemmi, Antonio started his offensive by writing letters to the New York press and to A. G. Bell, which remained unanswered. They also put together an article, signed by Antonio, stating that he was the first inventor of the telephone, inviting everybody to his cottage to see and test his different instruments and consult all the documents, including his 1871 caveat, which preceded Bell's patent by five years.

The response was of great solidarity. Almost all the newspapers published the article and sided with Antonio, who received a large number of letters expressing indignation. Strangers and neighbors alike would stop him in the street and encourage him not to give up fighting for his own recognition. While the Italian inventor was gaining unprecedented favor among the public, the National Bell Telephone Company became increasingly worried about the overly positive response of the press toward the Tuscan inventor, portrayed "as a poor immigrant, crushed by the arrogance and immense power of rich American industrialists."

It wasn't until nearly four years later that things started to improve for Antonio, thanks to a new telephone company. Between 1880 and 1885, several companies had been founded to commercialize the telephone, and one of these was the Globe Telephone Company, established in 1883 with a capital of $10 million and affiliates in Illinois, Kentucky, and Baltimore. Its mission was "to manufacture, sell, license and lease telegraphic, telephonic, and electric instruments and supplies, as well as to acquire and dispose of patents, patent rights and inventions relating thereto."

The Globe Telephone Company had soon hired Antonio as chief electrician with the intention of using his telephone and possibly acquiring his rights. The directors of the National Bell Telephone Company, which by then had split into American Bell and International Bell (1880), became even more alarmed when they found out that on September 22, 1883, Meucci had signed a preliminary contract with Globe, by which the new company had indeed acquired his rights.

A meeting was arranged between the lawyers of the two companies, thanks to the advice of Thomas Watson, Bell's faithful assistant and collaborator. They hoped to make an offer that was acceptable to Globe Telephone Company, thus eliminating a tough competitor. E. B. Welsh was sent as an intermediary. He met for over an hour with Antonio's lawyer, Lemmi, who showed the caveat and the chronological account of Antonio's invention.

Welsh carefully examined every document and expressed some perplexity over the statement that Edward B. Grant had kept Meucci's whole file, drawings, and instrument for over two years. He stated with great conviction that he had met Grant, who told him that all the material had been returned. It is indeed a mystery why the two of them had met and discussed Antonio's invention. In any case Grant, with his contradictory behavior, was certainly among the main people responsible for defrauding Antonio.

The Globe Telephone Company had no intention of negotiating because they felt secure and strong enough to sustain a real battle with their powerful rival, on the principle that A. G. Bell's patent had been fraudulently obtained. Moreover, in a democratic country like the U.S., Globe felt entitled to offer the general public an affordable service, an alternative to that provided by Bell in a healthy competition.

On the other hand, the American Bell Telephone Company was becoming increasingly concerned and tried very hard to avoid an open conflict. Their institution had been in existence for over six years and had accumulated enormous capital, while the Globe's new objective was very clear: "intensive growth, consolidation, and ramification of the financial activities and investments." The Globe was about to install more than 200 telephones in Baltimore. If this open competition were to be tolerated, even at the periphery of the empire, Bell's allies, such as Western Union, feared they would lose the 20 percent share of the profits they had secured through their agreement with American Bell.

For a short while Antonio enjoyed some recognition. In September 1884, his name was carved on a plaque at the main entrance of the Globe Telephone headquarters at 15 Broad Street in New York. During the same year his telephones (1857 and 1867 models) were defiantly and prominently exhibited at the International Electrical Exhibition in Philadelphia, along with that of Bell. He was given royalties of $200 a month, which alleviated his destitute condition.

Ester passed away a few months later, on December 21, 1884, finally at peace with herself and her husband, who, by the time of her death, seemed to have succeeded in gaining the long-overdue recognition they both knew he deserved.

However, by transferring his rights to the Globe Telephone Company for a symbolic amount, Antonio effectively lost control of the future of his invention. When the Globe refused two offers, one from Bell and the other from Edward Grant, Antonio was not consulted. As a courtesy he was informed—along with his lawyer, Michele Lemmi, and Angelo Bertolino—that offers had been made, though the amount was not specified. At that time William Goodwin, the president of Globe Telephone Company, warned Meucci and his representatives that a tough confrontation would be inevitable.[51]

In August 1885, two men—Watson Van Benthuysen of New Orleans and Charles P. Huntington of Mississippi—filed grave accusations of fraud against A. G. Bell to Justice H. W. McCorry of the U.S. District Court of Tennessee. They claimed that Bell was not the first and original inventor because at least Johann Philipp Reis, Antonio Meucci, and Elisha Gray had previously invented versions of the electric speaking telephone. The two accusers were asking to annul the two main Bell patents, vigorously maintaining that the American Bell Company defrauded the United States government, which had granted Bell's patents. Therefore they intended to request that a suit be instituted directly by the U.S. government against the American Bell Company and A. G. Bell in order to void his patents.[52]

The grave accusations were given the consideration they deserved. A formal investigation was opened at the Department of the Interior, while the American Bell Telephone Company unleashed a fierce campaign to block the proceedings, which yielded immediate results: Attorney General Augustus Hill Garland intervened directly and ordered the case closed. *The New York Times* openly attacked this unilateral decision and suggested he had a personal, direct interest in the American Bell Telephone Company affair. The controversy became highly political and violent, and other important public figures were scrutinized for possible conflicts of interest. To suppress the scandal, President Grover Cleveland stepped in and endorsed the government inquiry.[53]

As a result of this dispute and the subsequent media uproar, Antonio Meucci was spotlighted by the press, which unanimously sided with him and recognized him as the pioneer of the telephone.

At the end of October 1885, the Globe Telephone Company decided it was the right time to join forces with other companies, such as the Washington Telephone Company, to oppose the aggressive American Bell monopoly. While gathering evidence against American Bell, Globe's president, William Goodwin, found proof of disturbing irregularities, willful transgression, and widely spread corruption at the Patent Office. When he asked to see Meucci's caveat, he found it lying on a desk, against all regulations. The justification was that it had recently been shown and consulted. After assembling all the documents translated into English—the caveat, the drawings, as well as Meucci's precious Memorandum Book—and collecting a number of sworn statements in favor of Meucci's priority—the Globe Telephone Company was now ready for the ultimate clash.

They officially petitioned the U.S. government to sue the American Bell Telephone Company and "vacate the patents issued to Prof. Bell by reason of fraud, misrepresentation, mistake, and irregularities in the procuring of said patents."

The case was so serious that it was handled directly by Lucius Q. C. Lamar, Secretary of the Interior, the government department responsible for the Patent Office.

The first hearing took place in Washington on November 9, 1885, to investigate and determine whether there were sufficient grounds for a government case against Alexander Bell and the expanded American Bell Company. The Globe lawyers had the opportunity to show all of Meucci's documents—including a strong statement by Thomas Edison affirming that Bell could not be considered the first inventor of the telephone—and denounced the mysterious disappearance of Meucci's drawings and caveat itself from the files of the Patent Office.

The American Bell Company realized there was no time to waste and decided to counterattack and intervene right away. On November 10, 1885, the day after the first Interior Department hearing, American Bell filed an infringement suit against Globe Telephone. The infringement concerned Bell's patents and illegal use of apparatus, apparently very similar.

At this point there were two lawsuits pending and running parallel to each other: the U.S. government suit for fraud against American Bell in Washington, instigated by the petition of Globe Telephone, which started in March 1885; and the American Bell suit—on the initiative of one of the foremost patent attorneys, James J. Storrow, personal legal adviser to Bell—against Globe Telephone for patent infringement in New York as of December 1885, which lasted until the verdict on July 9, 1887.

The court case endorsed by Secretary of the Interior Lamar—the United States versus American Bell—was proceeding in

Washington. While the American Bell Company presented a 15,000-page defense, the Globe Telephone Company felt so sure of their victory that it proceeded to set up the Meucci Telephone Company in Elizabeth, New Jersey. Meucci received one hundred shares in the company and the assurance that a dignified tomb would be erected for his recently deceased wife. The new company started installing telephones according to Meucci's model on Globe systems as of March 24, 1886.

About a month later, on April 28, 1886, Bell and his father-in-law, Gardiner Hubbard, were called to testify under oath before the Committee appointed by the House of Representatives. The aim was to investigate the charges concerning the patent granted by the United States to Alexander Graham Bell, dated March 7, 1876, Patent No. 174,465, for improvements in telegraphy, and Patent No. 186,787, dated January 30, 1877.

The government suit declared that Bell's telephone patent application and claim were "purposely framed in ambiguous and general terms in order to cover both antecedent and future inventions, to deceive and mislead the examiners of the Patent Office as well as the public." Bell had never in fact been able to transmit articulate speech by the method or with the apparatus described in the application. He did not declare, the suit stated, that his alleged invention had any relation to the art of transmitting articulate speech by means of electricity, but entitled it an application for "an improvement in telegraphy": He referenced a then-recent patent application he himself had filed for a method of "multiple telegraphy," and treated his alleged new invention as another, more advantageous variation of the same contraption, but did not include, mention, or claim for it any capacity to transmit speech.

Bell was asked 156 questions. His replies were polite but evasive. When asked by Congressman W. C. Oates of Alabama if he was ready to prove he was the inventor, he answered, "I am ready to answer any question you think proper to ask." Neither the congressman nor anyone else on the committee pressed Bell for a straight answer. The improper nature of how Bell secured his first telephone patent in 1876 was the reason why the U.S. government pursued him for twelve years, even though since 1879 Bell had no connection with the company except as a stockholder. In fact, it was Gardiner Hubbard who had launched the patent application and had been the architect of the company's strategy from the beginning. The American Bell Company was doing everything in its power to delay this ongoing legal action.

The American Bell versus Globe-Meucci case was clearly a pretext, since the infringement regarded a single telephone line, the only one the Globe had already put up, from the company's office on Broad Street in New York City to another building, literally across the street.

For two years American Bell had tried insistently to find an agreement on their own terms without taking any direct action. Once on the defensive, they resorted to a countermove, hiring the best lawyers and detectives money could buy and instructing them to take action without hesitation.

William Goodwin of Globe Telephone—as well as the general public—knew very well that the suit initiated by

American Bell was a retaliation meant "as a means of intimidation and threat to prevent such an investigation." However, since they had been openly attacked, the Globe team had to defend themselves to establish the truth of the matter. Goodwin appointed attorney David H. Humphreys to defend Globe, Michele Lemmi was given the delicate task of coordinating the witnesses and defending Meucci's priority in the phone invention, while Angelo Bertolino was asked to help Meucci provide copies of all the models constructed after 1850 on Staten Island.

While American Bell had very influential politicians on their side, the Globe could count on the endorsement of the press. According to them, the Bell monopoly was hurting the U.S. economy by costing American taxpayers $10,000 a day. The Globe hoped that the press would persist in their battle without succumbing to intimidation. Its team wanted to be recognized as honest competitors to American Bell in a free trade economy.[54]

Before the actual legal actions between the Globe and American Bell took place in court, the American press unleashed an unprecedented campaign. *The Brooklyn Daily Eagle* suggested that all the patents granted to Bell were clearly obtained through fraud. *The New York Times* underlined how Bell's application concerned improvements on telegraphy, but the description and the drawings were so shrewdly and ambiguously crafted that they could apply not only to telegraphy, but also to the transmission of the human word. So, under the pretence of obtaining one thing, they obtained another, thanks to the incompetence or corruption of the Patent Office Examiner. Furthermore, the patent, once granted, was used to obtain the monopoly over an artifact that had been known much earlier, way before Bell presented his application. The government had been cheated and defrauded, damaging the people who had been deprived of their rights.

Moreover, the *New York World*, the *Baltimore Sun*, and *The Sentinel* in Chicago revealed disconcerting details about the telephone affair. It was pointed out that five years before Bell, Antonio Meucci had obtained a caveat in 1871 for a telephone that was being used on a daily basis in his cottage, but the drawings accompanying the caveat had disappeared from the Patent Office. Soon, strange as it seemed, the whole file had vanished and the only trace left at the federal office was its registration number.

All the major newspapers agreed that Meucci was the first inventor of the telephone and quite a number of people produced sworn statements declaring the telephone was in use on Staten Island and worked well before 1871. Across the ocean, the Austrian Ministry of Commerce and the Hungarian Ministry of Agriculture took the initiative to invalidate Bell's patents and rejected any offer of collaboration from the compromised American company.

After the various arguments were presented in public hearings at the Department of the Interior in Washington, the conclusions were all in favor of Meucci as the inventor of the first electric telephone.

Secretary Lamar strongly recommended to proceed on the basis of the evidence gathered, so that a lawsuit should be instituted; eventually it happened in March 1885 (*United States v. Bell Telephone Company*). Using the power of the Constitution, Article 1, Section 8—which, among other things, grants Congress the authority over issuing patents and copyrights to promote science and useful arts—the Justice Department presented a legal action against American Bell and Alexander Graham Bell in order to annul the patents granted on March 7, 1876, and January 30, 1877. The official bill against Bell included seven accusations:

1. Bell's elusive patent applications were written in ambiguous terms in order to deceive and disorient the Patent Office.
2. The above patents were granted illegally.
3. Bell had come to know the text of Elisha Gray's caveat through the federal Patent Office.
4. Gray was asked to abandon his project by giving up his caveat.
5. The undulating current was not invented by Bell but by Michael Faraday (and applied to the telephone by Wilhelm Weber and Werner von Siemens). Moreover, Bell was preceded in the invention of the electric phone by Reis, Varley, Meucci, Gray, Edison, Eaton, and others.
6. These different prototypes of phone were by then in the public domain, and therefore not patentable.
7. The patent granted on January 30, 1877, was also illegally and improperly procured.[55]

Attached to the recommendations to the Department of Justice were the actual bill, including the accusations against Bell; the transcripts of the various arguments heard; and over fifty affidavits in favor of Meucci. Zenas F. Wilber, the Patent Office Examiner, was responsible for five of them. In his final affidavit he claimed he intended with "clear mind" and "active conscience" to rectify as far as possible "any wrong" he "may have done." Although somewhat contradictory, all five affidavits charged that "A. G. Bell and his lawyers had engaged in fraudulent activity." The last confessional one, however, was not included in the official records, while others, discrediting Wilber, were allowed. We know of its existence thanks to *The New York Times*, which published the affidavit in its entirety on May 22, 1886.

In the government's report, Meucci's name was only peripherally mentioned along with early telephone inventors such as Reis, Varley, and Gray. It did not mention at all the disappearance of Antonio's file and prototype from Grant's office, and later on from the Patent Office, or any other offense against the Italian inventor. This lack of consideration for an Italian immigrant engaged in an almost thirty-year invention of the telephone would have a significant impact on the later developments of *American Bell Company v. Globe Telephone Company.*[56]

While the *United States v. Bell Telephone Company* case was dragging on, the *American Bell Company v. Globe Telephone Company* case began.

Meucci, now seventy-seven and with a long white beard, was called to appear on December 7, 1885, along with his interpreter Piero Mora, at the U.S. Circuit Court for the Southern District of New York. Chairing the debate was federal Judge William James Wallace, recently elevated to the U.S. Circuit Court for the Second Circuit, who had distinguished himself in the Civil War under President Abraham Lincoln. Meucci was defended by the Globe Telephone Company lawyer, David H. Humphreys, and questioned for over a month by James J. Storrow, one of the most feared patent lawyers in America, representing Bell and the American Bell Telephone Company.

Meucci's only wish was to see his rights as inventor defended and recognized before he died.

The same people who had been summoned by the Department of Justice in Washington to testify against Bell, Antonio Meucci *in primis,* now had to defend themselves in a grueling cross-examination that went on for thirty-six days.

Judge Wallace started by asking Bell why he had sued the Globe Telephone Company. The answer was very clear: He had heard of his competitor's intention to install 200 Meucci telephones in Baltimore, violating Bell's right to be the only one able to use his own Patent No. 174,465, granted on March 7, 1876.

Attorney Michele Lemmi then asked Bell if he thought he was the only one to have invented the phone, and he answered that he had no doubt. His company was at that time responsible for 200 miles of cables, 181,000 telephone sets and an average of over one million calls a day. Lemmi also asked what he thought of Meucci's telephone, and Bell said he had no opinion whatsoever. What he knew for sure was that his own telephones were all working well. And when Meucci's lawyer asked if he had known Edward B. Grant, vice president of American District Telegraph, who failed to return Meucci's documents, Storrow objected to the question. Bell, however, answered all the same, saying that he had met Grant through his father-in-law a long time before, when he was very young and not interested in the telephone.

Storrow, a rather cynical and ambitious lawyer, then asked Meucci to come forward for his deposition and tell the court which kinds of studies he had undertaken in Florence, why he left his native city, and when.

Meucci said he had studied at the Accademia delle Belle Arti and, after working as stage technician and engineer at the prestigious La Pergola theatre in his native town, left in 1835 with his wife for Cuba with a five-year contract from Teatro Tacón in Havana.

Wallace and Storrow already knew a lot about the defendant through the Pinkerton agency, the detectives American Bell had hired to uncover any negative information about Meucci. It turned out that Antonio was retired, old, and feeble by then, still living in the Staten Island cottage (which was no longer his), having lost a great deal of money while inventing his telephone but still hoping to recoup some of his investment. He had lost his wife in 1884 and had no children. He was indigent and lived the life of a recluse. Since his arrival in the States, he had never publicly exhibited his telephone, mainly because he didn't know any English and was constantly trying to improve his invention.

Meucci's neighbors described him as "one of those honest and honorable men who looked upon everyone as being like himself."

Since Antonio was so dignified and reserved, it had been difficult for the detectives to find out about his affairs except that he was receiving $200 per month in royalties and that his only revenue and support for his basic necessities came from the Italian Benevolent Societies of New York and Brooklyn. But it was well known that after his arrival from Havana—where he had been a successful inventor—he had become a friend of General Garibaldi, to whom he had given hospitality for eight months. Wallace and Storrow also knew that a girl had been living at the cottage for years, mainly to take care of Antonio's invalid wife, Ester, and had also often helped Meucci to test different phones. It was she who, on Ester's behalf, found a possible buyer for his telephone prototypes after the ferry boat accident.[57]

At the trial all the questions were asked in Italian and Meucci answered through his interpreter, which created multiple issues from the very beginning.

First, though he gave very prompt and lucid answers, his explanations often appeared unclear, due to Piero Mora's inability to properly translate technical terms into English. Antonio spoke at length about his caveat dated 1871 and expressed his conviction that this legal document, though expired, proved he was the true inventor of the phone, or at least the first one. Another strong point of his lengthy deposition was the Memorandum Book. It was a very precious record of the development in sequential stages of his major invention, where he had jotted down in pen, with further annotations in pencil, all of his experiments and prototypes from 1860 up to 1871, the year of his caveat registration.

While speaking, Antonio was sketching each apparatus he was referring to on a block of paper, showing his absolute mastery of the subject in every single answer. For example, this was his description of one of his early prototypes, dating back to 1853:

> *It is a tube of tin with the mouthpiece of the same material. Beneath the mouthpiece is an animal membrane saturated with bichromate of potash to render it firmer. In the interior is a small tongue of platinum soldered to a conductor of copper that communicates with the battery, and also in the interior is a tube with wire all around it, forming something like a bobbin. One extremity of the wire goes to the interior of the small tongue of platinum, the other to the battery.*

In spite of Meucci's convincing and precise arguments, Storrow started casting doubts on his technical preparation by poking him for his poor knowledge of the English language. Antonio would keep calm and kindly specify that he was speaking from experience while proceeding to describe at length the 1856–57 model, pointing out all the differences, mainly in the construction of the diaphragms, which at the beginning were of animal skin or different types of fabric, and later on of iron. He also lingered on the 1858–60 model, which used for the first time a bobbin that he had bought at Charles Chester's store in

New York City, his supplier.[58] Storrow's skeptical reaction was so obvious that Antonio started doubting whether the whole trial was a partisan machination to persistently discredit him in the attempt to finally get rid of "that presumptuous Italian immigrant," as the Bell team had defined him.

While describing the milestones of his invention, Meucci brought to the attention of both Humphreys and Storrow the fact that he had been asked in 1860 by the editor in chief of New York's Italian newspaper, *L'Eco d'Italia*, to write an article illustrating his telephone, which he did very willingly.

Storrow pretended to show a great interest in trying to locate the article, but it could not be found anywhere, neither at the newspaper editorial office, damaged by a fire, nor, oddly enough, in any of the most important public libraries in the entire United States: the New York Public Library, the Library of Congress in Washington, D.C., and the Center for Research Libraries in Chicago. All these libraries are missing the same numbers of the newspaper from 1853 to 1862—the very ones missing from the archives of *L'Eco d'Italia.* It appears that someone intentionally and illegally concealed or destroyed important judicial evidence of the period, thereby altering historical truth.[59]

In spite of a $100 reward offered to anyone who could bring in a copy, the mysteriously missing, removed, or stolen issue was never returned. (However, the issue of *L'Eco d'Italia* dated March 1880 was widely available—perhaps because, by then, Bell's patent had been successfully granted.)

All Meucci could do at this point was to reconstruct from memory the 1860 article published in *L'Eco d'Italia*, while the interpreter translated its brief but poignant text:

> *Antonio Meucci, who lives on Staten Island, has discovered a way of transmitting the spoken word by means of electricity, or with an electric conductor. He has carried out experiments on this subject for a long time and has obtained excellent results. His method consists in the use of two instruments, one for transmitting and one for receiving the spoken word. These instruments are quite easy to construct. Inside them is a bobbin of metallic wire. The center of the bobbin is a bar of tempered, strongly magnetized steel, and above it, is a diaphragm. The instrument, connected with a Bunsen or other type battery, transmits human speech exactly, just as it is pronounced by the two people who are in communication by means of insulated metallic wires.*

However, though the article was lucidly reconstructed, with the editor in chief of *L'Eco d'Italia*, Giacomo Secchi, in attendance, Storrow did not take this declaration into account, since the original printed copy of the newspaper could not be provided.

Antonio also mentioned that shortly after the article had been published, his friend Enrico Bandelari was about to leave for Italy, and he had promised to arouse interest there and possibly find Italian investors. Antonio had given him copies of his telettrofono drawings and a technical data sheet, and Bandelari had likely brought a copy of the article as well.

Meucci was also asked about his profession. He said he was an inventor who lived off the meager income from his patents, and that he had devised the first acoustic telephone in 1834 while working at Teatro della Pergola in Florence. He constructed a type of telephone to communicate between the stage and the control room, built on the principles of pipe telephones, later used on ships. In Havana, he explained, he invented a more advanced apparatus, thanks to the electrotherapy he was practicing. Storrow grabbed this occasion to insinuate that Meucci had actually built the instrument mainly to alleviate physical pain rather than to transmit the human word. The audience started booing, vehemently reacting to Storrow's disrespectful insinuations.

Though Antonio specified that, thanks to the generosity of his close friend Lorenzo Salvi, he was able to study Mesmer's theories and Becquerel and Galvani's treatises, Bell's lawyers did not seem to take his scientific competence seriously. Antonio underlined that when he was actually experimenting with his first electric telephone in Havana on New Year's Eve in 1849, Bell was only two years old, living in Edinburgh.

He explained that he moved to the States with his wife in 1850, settled on Staten Island, began conducting experiments regularly, and that Garibaldi had the opportunity of seeing his work. Storrow immediately dismissed Garibaldi's appreciation because he was a general and not a technician.

Antonio also recounted his terrible accident during the ferry boat explosion in 1871 and his wife's unfortunate decision to sell his prototypes to junk dealer John Fleming, out of dire necessity. He recounted that when realizing that no trace of his major scientific project remained in his lab after the ferry explosion, he felt the need to apply for a caveat in order to protect his main invention. This account offered Wallace and Storrow the irresistible opportunity to state that if even Meucci's wife was ready to dispose of his prototypes and clear out his laboratory, they most likely could be considered of little value, if not worthless.

Numerous witnesses and friends came to demonstrate their solidarity with Meucci, each of them with his or her own part of the story to tell.

Fleming came with a number of spools of copper wire, reels, and diaphragms in the shape of a horseshoe, four inches long and three inches across, that were left behind in his shop. He stated that Mrs. Meucci had sold all of Antonio's equipment to him, which he then resold to an unidentified gentleman from New York. In a sworn affidavit, Fleming specified that all these items were indeed bought in 1871 after the ferry boat accident.

Angelo Bertolino, Meucci's faithful notary public, recounted in detail their visits to Globe Telephone Company's Edward B. Grant and his ambiguous behavior until—after keeping Antonio in suspense for almost three years—he maintained he had misplaced his file, including the drawings and telephone set.

Matilde Ciucci, who had lived with the Meuccis for a number of years to look after Signora Ester, recounted being constantly asked to go either upstairs or down to the basement to test the phone Antonio had installed on three different floors of the house, adding that she could hear every single word he and his wife would exchange.

Then there was Nestore Corradi, the painter who, in 1857, had made a drawing of two people calling each other over a transmitting and receiving instrument of the kind Meucci had installed in his cottage. The artist brought a copy of the drawing to court, since the original, enclosed with Meucci's caveat, had been stolen from the Patent Office.

Domenico Mariani, who had been a close family friend for over twenty-five years and had lived on and off in the Staten Island cottage, testified that he used the phone daily while he was there. Though now eighty-one, he eagerly provided a fifty-two-page deposition, including meaningful information about Meucci's frequent trips to New York to buy material for his experiments. He also described the perfectly functioning telephone lines installed at home.

Ferdinando de Luca, the Consul General of Italy in New York in 1867–68, submitted a written statement of great appreciation of Meucci while posted in China. Well-known physicist Ashael K. Eaton also wrote an important statement about Antonio's phone, firmly stating that it preceded Bell's (this document, however, was eventually left out of the dossier).

Enrico Bandelari, the friend who had volunteered to look for possible sponsors for Antonio's new invention while in Italy in 1860, also came to court, as did Charles Chester, owner of the New York shop where Antonio used to buy all he needed for his experiments. Also making an appearance were Enrico King, the family doctor who treated Antonio during his long convalescence, and patiently assisted him while he was in a state of despair after all his prototypes had been sold without his consent; Mattia Egloff, the tenor from Clifton who often helped Meucci in his experiments and sang "La Marseillaise" several times through the telettrofono installed at the cottage; and Giacomo Secchi, who asked Meucci to write the article about his "speaking machine" in *L'Eco d'Italia.*

William Goodwin of Globe Telephone was also called to testify. When asked why he contacted Meucci rather than Bell, he answered that he thought Meucci was the real pioneer of the phone. His intention, along with his associates, was to make a profit in an open market in competition with American Bell.

Storrow was a formidably capable lawyer and had been hired to win the case at all costs. He was competent in scientific matters, so he was very well informed and impressed the audience by using technical terminology. He incessantly tried to discredit Meucci's Memorandum Book, stating it was fake and backdated.

He was, however, contradicted by one of his own witnesses, William Rider, who had been Antonio's partner in the candle and paper factories for years before they went bankrupt. Despite this, Rider testified in favor of Bell. In a moment of honesty, he admitted that he himself had given a new ledger of the Rider & Clark Company to Antonio, since his Italian partner had a worn-out notebook full of drawings and loose papers. He also stated that the Italian inventor had "great moral integrity" and was "the last man in the world who would take anything that did not belong to him." Throughout his testimony, Rider pretended he didn't remember having seen the telettrofono at Antonio's cottage, but after the audience booed at him several times, he suddenly admitted he had heard Meucci's invalid wife talk from her bedroom to her husband in the lab down in the basement. But when he tried the instrument, he said, he heard only indistinct noises.

In spite of his obvious lies, Rider was fundamental in proving that in 1865 Antonio was already regularly experimenting with the phone—six years before submitting his caveat and eleven years before Bell's patent application. Ironically, it was Rider who provided the funds for Antonio to construct the first fine electric telephone with "excellent results in the transmission of the exact word."

After this last deposition, Storrow quickly decided to change his strategy. Despite persistent and excruciating cross-examination of Meucci's witnesses, he was unable to extract a single word against the defendant. So he started to disseminate doubts on everything the Italian inventor was saying and kept accusing him, with a touch of scorn and racism, of being self-taught and rather ignorant.

The testimony of Thomas Stetson, the patent attorney whom Antonio had hired to apply for his caveat, marked the turning point in the trial. Stetson admitted he had not entirely understood his client's invention when he submitted the application on his behalf, and therefore had neglected to integrate the corrections in pencil that Meucci had added in the margin. One of these was a clarification that the wires had to be covered with insulating material so that the users of his instrument would not receive an electric shock, as had happened previously in Cuba. As specified in his affidavit, he did not understand the complexity of the description because the caveat had not been accurately translated. In addition, he found the drawings unclear. He did not take the necessary time to clarify the details, since Meucci had not been able to put together the $250 necessary to apply for a patent and, at that time, he could only spend $10. Stetson decided to submit the caveat without Meucci's corrections.

The whole Stetson deposition was ambiguous and critical, to say the least. In the meantime, he had become a consultant for American Bell, as he was forced to admit when asked under oath. So, while Stetson formally appeared as a witness for Globe, he was substantially obliged to the opposing party.[60]

After agreeing with Stetson's remarks, Storrow introduced George F. Durant, who had since become director general of American Bell Telephone in Missouri. He was the person in the Western Union office who had been asked by Edward Grant to inform Meucci that his file and telephone prototype had been lost or misplaced.

When Antonio's lawyer, Humphreys, asked Durant why they waited so many years to inform Meucci, he evasively replied that he did not know. The American District Telegraph Company was only interested in telegraphy, yet they had reviewed Meucci's project and found it in a rough preliminary state. The drawings were approximate and the animal membrane of the diaphragm would have to be clearly substituted with a metal one.

Humphreys asked whether Grant had promised Meucci and Bertolino to do a few tests at his company. Durant admitted that, yes, it was indeed true, but the tests were never done because in the meantime Grant had become very busy and realized it was not worth the effort. Humphreys asked whether Bell would spend a lot of time researching there. Durant's answer was very evasive, yet he specified that certainly Bell and Grant were on very good terms.

Meucci found himself trapped in a most prolonged, biased, and humiliating trial, with the Bell team's primary objective to prove that Antonio had not invented the electric telephone, but rather a merely acoustic or mechanical instrument. Throughout the thirty-six days of Antonio's interrogation, his answers were always sharp and to the point. However, the trial turned out to follow a very well-rehearsed script, where questions and answers had already been decided.

Bell and his powerful, committed lawyers decided to state their case showing once and for all that he was the sole inventor of the electromagnetic phone. Among the witnesses called to support Bell's predetermined thesis were Professor Charles Robert Cross, a personal friend of Alexander Bell meant to be the star witness, and Frank Pope, the head engineer of Western Union when Gray and Bell were experimenting there. Pope had distinguished himself by first attributing the invention of the telephone to Elisha Gray, only to suddenly change his mind as soon as he started working for the American Bell Telephone Company.

So when Humphreys asked Pope what he thought of Meucci's telephone, he adamantly answered that he did not consider it a phone. He maintained that the way the caveat had been written did not convey the idea of an instrument capable of transmitting the human voice by means of electricity, but rather through mechanical vibrations.

Humphreys took the opportunity to ask him about the experiment he conducted at home with his brother Henry, by then the director general of the American District Telegraph Company, using two phone sets from the warehouse. Pope admitted that they had managed to talk for a few hours between their two nearby houses. When Humphreys asked who had designed those telephones, Pope was very evasive and stated that he never found out.

Charles Robert Cross, a highly esteemed professor of physics at the Massachusetts Institute of Technology, had been hired with the objective of helping Bell win the case without any scruples or ethical restraint. He was chosen to lead the offensive as he was a master of confusion, avoiding confrontation, and pompously using erudite jargon. He too insisted that Meucci did not invent an electric phone but simply a gadget, a lover's telephone with which he had been playing for many years.

Humphreys invited the jury to visit Meucci's cottage and see his electromagnetic telephone in action, insisting that the most important thing was to verify that his telephone was working, but the invitation was promptly declined. Cross kept repeating that it worked well because it was copied. His role on the stand was to conclude in a very contorted way that Meucci was only able to create, as best as he could, a mechanical instrument. He said that Antonio's reputation, on the other hand, was that of a sorcerer's apprentice of sorts, a kind of Italian magician, because of his strange experiments in his cottage known to the neighbors for years as the "devil's house."

Cross repeatedly stated that Meucci's caveat application was not clear. Both the description of the apparatus and the drawings attached were not precise at all. The professor's words carried weight, given his prestigious position at the university. Storrow gave the impression of slowly coming to appreciate his erudite remarks.

A brilliant and cunning patent attorney, Storrow was a master of sophistry. He had calculated every move. He envisioned from the very beginning how he and Cross would work together to win the case, simply by resorting to subtle defamation and trivialization of the real inventor of the electromagnetic phone and ultimately treating him like an impostor.[61]

The team worked in the single direction of discrediting the Florentine inventor, portraying him as a self-taught person of such limited learning that he could never have invented such a sophisticated instrument as the electric phone. While Storrow kept pontificating, Cross persistently underlined that the 1871 caveat made no mention of magnet, coils, or diaphragm and, given Meucci's limited scientific background, it was simply a string telephone, or a mechanical one at the most, consisting of a mouthpiece and an earpiece connected by a wire, and that "beyond this, the invention of Meucci was only imagination."

At the conclusion of the trial, upon Storrow's request, a recent article was added to the whole dossier. It described Meucci as a candle maker who had constructed an acoustic telephone believing it was electric. The Bell team conveniently ignored Antonio's Memorandum Book, which contained all the details of his experiments with different prototypes and their materials, claiming it was a forgery. Meucci's reputation was compromised to the point that, about a hundred years later, Pulitzer Prize–winning writer Robert V. Bruce had the audacity to agree with Storrow and describe Antonio as "the silliest and weakest impostor who has ever turned up" in his 1973 biography, *Bell: Alexander Graham Bell and the Conquest of Solitude.*[62]

Meucci was recalled for additional clarifications in September 1886 by the Circuit Court of New York to defend himself and the Globe against the charges of patent infringement initiated by American Bell. In response, he assembled a 214-page deposition that included his Memorandum Book, translated into English by Michele Lemmi; the drawings and models; as well as the affidavits from his witnesses. He forwarded the dossier to Judge William James Wallace, who, in the past, had decided in favor of the Bell Company against four other claimants.

Judge Wallace was a native of Syracuse, New York, a Republican who had served as mayor of his city before being appointed federal judge for the Northern District of New York. In 1882 he was further promoted to the U.S. Circuit Courts for the Second Circuit. He studied the trial records and all the documents for almost ten months until issuing his final verdict on July 19, 1887.

Wallace's judgment against the Globe Telephone Company and Meucci was entirely in favor of Bell. He agreed with Cross and Pope and concluded that Meucci's 1871 caveat did "not describe any of the elements of an electric speaking phone" but rather "a mechanical telephone." He insisted it was due to the insufficient scientific background of the Italian inventor. Besides pointing out how the sketches Meucci had made during the trial were not clear at all, Wallace insisted on the equivocal statement in the caveat concerning insulation. He even proceeded to discredit the Telettrofono Company, alleging that it had been founded "on the credulity of his three friends, who had small means and little business experience." His derogatory conclusion was that Meucci himself knew that "his invention was without any real value because he had not succeeded or even tried to develop it commercially."

According to Giovanni E. Schiavo, a very accurate biographer of Meucci, the verdict was "one of the most dishonest legal decisions in the annals of American justice, and not only dishonest but also outrageously offensive, a complete fabrication of lies and distortions."[63]

Why did Judge Wallace come to this conclusion that left Meucci, the Globe Telephone Company, and most of the press literally speechless? Was he prejudiced against Italian immigrants? Was he misled by the famous patent lawyer Storrow or the erudite considerations of Professor Cross?

It seems to be the case, as others also maintain, that it was above all a matter of defending the national economy. American Bell had quickly built an empire spanning North America and

Europe, propelled by high capital investments. In 1887 alone the American Bell Telephone Company invested almost $60 million throughout the United States, establishing telephone lines connecting main cities, a true revolution in the communication system. Though the company had spent nearly $1 million in legal fees on controversies in its first ten years of life, in 1882 American Bell Telephone had earned its first $1 million, and by 1884 had paid its first $1 million in dividends.[64]

A judgment against such a large and significant company could well disturb the American economy, even though the legality of their patents was indeed debatable. If the American Bell Company had lost the case, it would have been economically devastating as well as catastrophic for the factories that manufactured the telephonic instruments. They had reached a level of growth that represented an example of a young, vibrant, progressive, united American nation projecting positive messages of economic efficiency. Perhaps Wallace also believed that such a major instrument as the telephone, which had changed world communications, should be invented by a cultured Anglo-Saxon rather than a poor Italian immigrant.

Judge Wallace's decision, a grave miscarriage of justice, was issued at the end of July 1887, two years before Meucci's death. It was appealed to the Supreme Court of the United States in October 1888 by the Globe Telephone Company and Meucci. The company had to stop any expansion and further investment as a consequence. They also suspended the agreed payment of royalties to Meucci, and even had to bear the cost of the appeal.

In a state of despair, Antonio wrote his last letter to his two brothers, Giuseppe and Luigi, informing them that he was still plagued by poverty. When the case was finally called for hearing in October 1891, the Globe failed to press the appeal because in the meantime Meucci had died, on October 18, 1889. The company had already invested all of its hopes in the much more important *U.S. v. Bell* trial, which was still pending. Consequently, the appeal was dismissed on March 10, 1892.

The progress of the *U.S. v. Bell* trial, launched in 1885, was interminably slow. The Bell lawyers, by a well-planned policy of procrastination and obstructionism, dragged the case on for twelve years to November 30, 1897—eight years after Meucci's death—without ever resolving the key question of who had legal claim to the invention of the telephone. The case was simply abandoned and no decision was ever taken. As a consequence of dropping the case, the U.S. government couldn't claim that Bell had not invented the telephone, nor could the Bell Company claim that he had. All the same, A. G. Bell continued to be credited with the invention of the telephone, with no mention of Meucci or anyone else.

One hundred and thirteen years after Antonio Meucci's death, the U.S. Congress decreed in 2002 that Meucci's discoveries were ahead of his time by as much as thirty years, and therefore concluded that Meucci had invented the telephone. Finally, this unique Italian inventor was given his rightful place in the history of science.

Epilogue

The history of the telephone will never be fully written. It is partly hidden away in twenty or thirty thousand pages of testimony and partly lying in the hearts and consciences of a few whose lips are sealed, some sealed in death and others by a golden clasp whose grip is even tighter.

—Elisha Gray

Antonio Meucci was obsessed with the transmission of the human word throughout his adult life, since his first major assignment in Florence at La Pergola theatre. It was only after almost thirty years of research that was he able to perfect his invention and apply for a caveat. A. G. Bell, though generally portrayed as a gentleman of integrity experimenting in telegraphy, applied in 1876 for the patent of an invention that was still just a coveted goal.

I am convinced that Alec Bell—as Mabel used to call him after they became engaged— never would have succeeded in receiving a patent for a telephone that he had not yet invented if he had been left alone to decide about his own life. Instead, he was overwhelmed and almost crushed by the strong, cynical, and persistently focused personality of Gardiner G. Hubbard, especially after the latter realized there was a possibility of having him as a son-in-law. A powerful and wealthy Machiavellian patent lawyer, Hubbard was highly regarded in the elite circles of Boston and Washington, D.C., and showed the future member of his family that nothing was impossible for him with the endorsement of his daring and unscrupulous partners.[65]

Bell's misbehavior in the telephone's invention occurred when he was still in his twenties. Whatever ethical compromises he may have made in his dealings with the Washington lawyers, or with the examiner of the federal Patent Office, he had most likely made out of love in order to overcome the broad gap between Mabel's economic standing and his own. As he wrote to Mabel one year after their wedding, in 1878:

> *Why should it matter to the world who invented the telephone so long as the world gets the benefit of it? Why should it matter to me what the world says upon the subject so long as I have obtained the object for which I laboured and have got you my sweet darling wife?*

He naively thought he could ignore the telephone monopoly plan of his greedy father-in-law and live in peace ever after. Instead, he was to spend a life of economic leisure and impunity, however besieged by over 600 legal battles that kept raging over time. Interestingly enough, Bell was cited as saying that he would rather be remembered as a teacher of the deaf than as the disputed inventor of the telephone.

Antonio Meucci, on the other hand, identified his whole life with the transmission of the human voice, and the reason he waited so long before applying for a provisory patent, or caveat, was due to the fact that he was a perfectionist and therefore wanted to count on an instrument that would indeed perform according to his high expectations. His passionate, relentless research was definitely slowed down by his increasingly sick wife, who had visibly lost interest in her husband's most expensive research. And unlike Bell, who was affluent and well connected, Meucci had to make a living within an unsympathetic social context, which relegated immigrants to an underclass of ruthless exploitation.

When Meucci died on October 18, 1889, at age eighty-one, he was of the firm conviction that his appeal to the Supreme Court would be successful. He never doubted that the priority of his claim would be recognized sooner or later. His faith was as solid as a rock.

In the last years of his life, cared for by the Nisini family, he was comforted by the good news that he had been assigned a modest yearly sum from the Italian Freemasonry of New York. Moreover, thanks to Garibaldi's daughter Teresita, he was also granted a small pension by Prime Minister Francesco Crispi, one of the main protagonists of the Risorgimento, on behalf of the Italian government.

When it became clear that Meucci's illness might be fatal, Crispi, who was also a close supporter of Garibaldi and Mazzini, cabled Giovanni Paolo Riva, the Italian Consul General of New York, informing him that the Italian government would cover all the medical expenses and a dignified funeral for the Florentine inventor.

Several American newspapers reported Meucci's death. In particular, the *New York Herald* gave a detailed account of his funeral:

> *The center of attention on Staten Island yesterday was the famous "Casa Garibaldi" (Garibaldi's homestead) and there at an early hour were gathered many hundred devoted Italians ready to accompany to their last resting place the remains of Antonio Meucci, the friend and business associate of Italy's great patriot, and no mean patriot himself. The quaint old house was thronged with visitors at nine o'clock. In a darkened room on the lower floor rested the plain black coffin, crowned with flowers, and by it passed in procession the friends and the representatives of many Italian societies, who had known and honored Meucci for many long years. Chief among the mourners were Messrs. Oldrini, Cadicamo, Barsotti, Contencin, Cantoni, Lemmi, Barattoni, Bonanno, Bachmann and Dr Abruzzo . . . Other prominent Italians, all in handsome uniforms represented the Garibaldi, Operaia and Mazzini societies in this city as well as from Stapleton, Tompkinsville, Clifton, Jersey City, Brooklyn, Newark and Paterson. Many of these delegates bore banners and were attended by bands of musicians . . . The remains were borne to a ferryboat chartered for the occasion to bring them to this city on their way to the crematory at Fresh Pond, Long Island. The*

representatives of the civil and military societies marched at the head of the procession and were followed by carriages containing the representatives of the Italian and English press and friends of the deceased.

The Staten Island cottage where Meucci lived for thirty-nine years was a gift of Frederick Bachmann, a German immigrant, who allowed Antonio and Ester, after their economic downfall, to stay for the rest of their lives without paying any rent. Bachmann's exemplary, benevolent attitude and increasing generosity toward the Italian inventor, as well as to the Italian community, stemmed from a deeply felt solidarity among uprooted Europeans transplanted in America.

After Meucci's death, the historic white cottage was transferred to 420 Tompkins Avenue in Rosebank—a few blocks from its original location—dismantled and then reconstructed exactly as it was. Today it is home to the Garibaldi-Meucci Museum, run and maintained in an exemplary way by the Sons of Italy Foundation. In the Garibaldi Room, the museum's main attraction, all of the general's belongings are still there, including his hand-carved walking stick, two rifles, and a deer antler he received in South America as a talisman. The cottage also contains a celesta that Meucci built, along with some of his inventions and other artifacts.

Until the early 1900s, however, Meucci's name evoked not only pride in the invention of the telephone, but also a deeply hurtful and humiliating defeat. Many Italian Americans lost faith in the judicial process of their new nation. Yet in 1922, more than thirty years after Meucci's death, Fiorello La Guardia, mayor of New York, described Meucci as "an Italian son that America will never forget."

In 1923 a monument to Meucci was installed at his cottage. It was sculpted in Italy by Ettore Ferrari, using bronze obtained from the cannons used by the Austrian army in the Battle of Vittorio Veneto in World War I. The Italian community was so grateful to the Italian inventor that in 1989 they dedicated a square to him in Brooklyn, marked by a stone memorial with the inscription "Antonio Meucci, 1808-1889, Father of the Telephone, First U.S. Patent Caveat 3335, Dec. 28, 1871."

Antonio Meucci's faith was finally rewarded 113 years after his death, when he was given credit as the true inventor of the telephone by the U.S. Congress in its Resolution 269, introduced by Congressman Vito Fossella and approved by the House of Representatives on June 11, 2002. Ultimately, this unique Italian immigrant, whose career was both extraordinary and tragic, was given his rightful place in the history of science.

Appendix

Hold the Line: the Life, Loves and Inventions of Antonio Meucci, Father of the Telephone is meant as a tribute to Meucci's scientific achievements, as well as to all those whose contributions to science continue to be neglected.

As Meucci biographer Basilio Catania pointed out on several occasions, for a long time, the Bell Company publicized the New York Court decision, which ruled in its favor. The records of this trial, in particular Meucci's deposition, are available in print in many public libraries. Therefore, anyone can read how the Bell lawyers defeated the Globe Company and Antonio Meucci. Before the Bell Company sued the Globe Company and Meucci for patent infringement, however, the U.S. government had filed a suit against the Bell Company and Alexander Graham Bell for fraud, collusion, and deception. The Bell team, however, neglected to say that.

In other words, the U.S. government was on Meucci's side, as it sought to prove that he, not Bell, had invented the electromagnetic telephone.

The action of the government, hampered by the obstructionism of the Bell lawyers, dragged on for twelve years, up to the end of 1897 (eight years past Meucci's death), without resolving the underlying question of who had priority over the invention of the telephone. Moreover, the record of this trial was never printed and is only now available, with difficulty, from the National Archives, mostly in typescript or manuscript, and spread among different record groups and cities.

On September 25, 2001, during the 107th session of U.S. Congress, the following resolution was made:

107th Congress, 1st Session, H. Res. 269

Expressing the sense of the House of Representatives, U.S., to honor the life and achievements of 19th-century Italian American inventor Antonio Meucci and his work in the invention of the telephone.

Whereas Antonio Meucci, the great Italian inventor, had a career that was both extraordinary and tragic;

Whereas, upon immigrating to New York, Meucci continued to work with ceaseless vigor on a project he had begun in Havana, Cuba, an invention he later called the "telettrofono," involving electronic communications;

Whereas Meucci set up a rudimentary communication link in his Staten Island home that connected the basement with the first floor, and later, when his wife began to suffer from crippling arthritis, he created a permanent link between his lab and his wife's second floor bedroom;

Whereas, having exhausted most of his life's savings in pursuing his work, Meucci was unable to commercialize his invention, though he demonstrated his invention in 1860 and had a description of it published in New York's Italian language newspaper;

Whereas Meucci never learned English well enough to navigate the complex American business community;

Whereas Meucci was unable to raise sufficient funds to pay his way through the patent application process, and thus had to settle for a caveat, a one year renewable notice of an impending patent, which was first filed on December 28, 1871;

Whereas Meucci later learned that the Western Union affiliate laboratory reportedly lost his working models, and Meucci, who at this point was living on public assistance, was unable to renew the caveat after 1874;

Whereas in March 1876, Alexander Graham Bell, who conducted experiments in the same laboratory where Meucci's materials had been stored, was granted a patent and was thereafter credited with inventing the telephone;

Whereas on January 13, 1887, the Government of the United States moved to annul the patent issued to Bell on the grounds of fraud and misrepresentation, a case that the Supreme Court found viable and remanded for trial;

Whereas Meucci died in October 1889, the Bell patent expired in January 1893, and the case was discontinued as moot without ever reaching the underlying issue of the true inventor of the telephone entitled to the patent; and

Whereas if Meucci had been able to pay the $10 fee to maintain the caveat after 1874, no patent could have been issued to Bell: Now, therefore, be it resolved, that it is the sense of the House of Representatives that the life and achievements of Antonio Meucci should be recognized, and his work in the invention of the telephone should be acknowledged.

September 25, 2001 (10:41 AM)

In the summer of 2022, during renovations of La Pergola theatre in Florence, a third acoustic tube was found. It would be remarkable to see it restored from a museum artifact into a functioning device that could be experienced during the frequent educational visits to this unique historical landmark.

Notes

INTRODUCTION

1. Seth Shulman, *The Telephone Gambit: Chasing Alexander Graham Bell's Secret* (New York: W. W. Norton & Co., 2008), 212.

CHAPTER ONE: THE YEARS IN FLORENCE/ FLORENTINE EDUCATION (1808–1835)

2. Franco Capelvenere, *Meucci. L'uomo che ha inventato il telefono* (Florence: Vallecchi, 2003), 22–23.
3. Franco Angotti, Giuseppe Pelosi, *Antonio Meucci e la città di Firenze* (Florence: Firenze University Press, 2009), 19.
4. Ibid., 48–49.
5. Ibid., 161.
6. Sandra Meucci, *Antonio and the Electric Scream* (Boston: Brandon Books, 2010), 30–31.
7. Giorgio Fantoni, *L'evoluzione del telefono* (Lecce: Youcanprint, 2019), 15.

8. Susan E. Hamen, *Who invented the telephone?* (Minneapolis: Lerner Publications, 2018), 13.

CHAPTER TWO: INVENTING IN HAVANA (1835–1850)

9. Basilio Catania, *Antonio Meucci: L'inventore e il Suo Tempo. New York 1850–1871*, vol. II (Turin: Seat, 1996), 212–218.
10. Sandra Meucci, *Antonio and the Electric Scream: The Man Who Invented the Telephone.* (Boston: Branden Books, 2010), 49–50.
11. Russell A. Pizer, *The Tangled Web of Patent #174465*, (Bloomington, Indiana: AuthorHouse, 2009), 221.
12. A newspaper of Baltimore, *The Sun*, published an obituary notice on Antonio Meucci on October 19, 1889, stating as follows: "In 1850, Meucci came to New York from Cuba, where his only child, a girl of 6, Juanita, had just died of yellow fever . . . ."
13. Catania, *Antonio Meucci: L'inventore e il suo tempo. New York 1850–1871*, 275.
14. Basilio Catania, *Antonio Meucci: Telephone Pioneer* (*Bulletin of Science, Technology and Society*, 21:1, February 2001), 55.
15. Marco Nese, Francesco Nicotra, *Antonio Meucci, 1808–1889* (Rome: Italy Magazine Publisher, 1989), 21.

CHAPTER THREE: STATEN ISLAND, NEW YORK (1850–1871)

16. Ibid., 28–29.
17. Felice Foresti was in prison with Silvio Pellico, the hero of the Italian Risorgimento, who wrote *Le mie prigioni* (*My Prisons*)

on the challenging twenty years he spent in jail, first at the Piombi at Doge's Palace in Venice and then in Špilberk in South Moravia.

18. Nese, *Antonio Meucci, 1808–1889*, 45.
19. Magnasco had become an American citizen, which allowed him to be appointed flag captain of the *Commonwealth*, the American merchant navy ship that would later take Garibaldi from Baltimore back to Italy in 1854.
20. U.S. Patent Office: Manufacture of Candles. Patent No. 22,789, January 25, 1859.
21. Catania, *Antonio Meucci: L'inventore e il suo tempo. New York 1850–1871*, 68.

    On September 10, 1851, Salvi bought Lot 1 of 2500 m^2 for \$1,800, south of the cottage for the candle factory to be erected.

    On May 12, 1852, Salvi and Meucci bought Lot 2 of 1000 m^2, paying \$4,000 (on a 50/50 basis), of which \$1,500 was a mortgage to the owners, on which the cottage stood.

    On July 1, 1853, they bought Lot 3 (again on a 50/50 basis) of about 5,500 m^2 for \$4,800, of which \$3,300 was a mortgage to the owners).
22. In 1854 he would be asked by the general to accompany him to Italy on the *Commonwealth* because he had attended the Naval Academy in Venice and could be of great help.
23. Catania, *Antonio Meucci: L'inventore e il suo tempo. New York 1850–1871*, 121.
24. Sandra Meucci, *Antonio and the Electric Scream*, 70.
25. Ibid., 168–169.

26. Nese, *Antonio Meucci, 1808–1889*, 65.
27. Catania, *Antonio Meucci: L'inventore e il suo tempo. New York 1850–1871*, 157–158.
28. Nese, *Antonio Meucci, 1808–1889*, 71.

CHAPTER FOUR: STATEN ISLAND, NEW YORK (1871–1889)

29. Catania, *Antonio Meucci: L'inventore e il suo tempo. New York 1850–1871*, 258.
30. Nese, *Antonio Meucci, 1808–1889*, 93.
31. Basilio Catania, *The Meucci-Bell Dichotomy* (Milan: R&D Mond, 2001), 18.
32. Shulman, *The Telephone Gambit*, 18–20.
33. Ibid., 32–34.
34. George Bernard Shaw, *Pygmalion* (London: Penguin Classics, 2003), preface.
35. Catania, *The Meucci-Bell Dichotomy*, 35.
36. Nese, *Antonio Meucci, 1808–1889*, 97–99.
37. Shulman, *The Telephone Gambit*, 67–73.
38. Ibid., 93–95.
39. Thomas Watson, *Exploring Life: The Autobiography of Thomas Watson* (New York: D. Appleton and Company, 1926), 66–70.
40. Pizer, *The Tangled Web of Patent #174465*, 5.
41. Ibid., 143.
42. Ibid., ix.
43. Ibid., 55–56.
44. Shulman, *The Telephone Gambit*, 171–172.
45. Pizer, *The Tangled Web of Patent #174465*, 67–69.

CHAPTER FIVE: ANTONIO FIGHTS BACK (1876–1889)

46. Nese, *Antonio Meucci, 1808–1889*, 103–107.
47. Catania, *The Meucci-Bell Dichotomy*, pp. 32–33.
48. Russell A. Pizer, *The Tangled Web of Patent #174465*, p. 10.
49. Matthew Josephson, *The Robber Barons* (New York: Harcourt, Brace and Company, 1934), 383.
50. Paul Latzke, *A Fight with an Octopus: Being the Story of a Great Contest that Was Won Against Tremendous Odds*, As Printed Originally in *Success Magazine*, Appendix H, 1906.
51. Capelvenere, Meucci. *L'uomo che ha inventato il telefono*, 173.
52. Nese, *Antonio Meucci, 1808–1889*, 129.
53. Ibid., 131.
54. Capelvenere, *Meucci. L'uomo che ha inventato il telefono*, 184.
55. Ibid., 192.
56. Catania, *The Meucci-Bell Dichotomy*, 35–37.
57. Pizer, *The Tangled Web of Patent #174465*, 281–301.
58. Nese, *Antonio Meucci, 1808–1889*, 139.
59. Catania, *Antonio Meucci: L'inventore e il suo tempo. New York 1850–1871*, 139.
60. Catania, *The Meucci-Bell Dichotomy*, 34.
61. Capelvenere, *Meucci. L'uomo che ha inventato il telefono*, 218–219.
62. Robert V. Bruce, *Bell: Alexander Graham Bell and the Conquest of Solitude* (Boston: Little, Brown, 1973), 272.
63. Giovanni E. Schiavo, *Antonio Meucci: Inventor of the Telephone* (New York: The Vigo Press, 1958), 102.
64. Nese, *Antonio Meucci, 1808–1889*, 159.

EPILOGUE

65. Pitzer, *The Tangled Web of Patent #174465*, 244–245.

Bibliography

Aitken, William. *Who Invented the Telephone?* London and Glasgow: Blackie and Son, 1939.

Angotti, Franco and Giuseppe Pelosi. *Antonio Meucci e la città di Firenze.* Florence: Firenze University Press, 2009.

Angotti, Franco and Giuseppe Pelosi. *Il telefono e dintorni.* Florence: Firenze University Press, 2011.

Antonio Meucci, Cittadino Toscano, Contro il Monopolio Bell. Italy: RAI Radiotelevisione Italiana, 1970.

Antonio Meucci: The Father of the Telephone. New York: De Nonno Productions Inc., 1991.

Barrett, W. F. "The Telephone. Its History and Its Recent Improvements." *Nature*, November 7, 14, and 21, 1878.

Barzini, Luigi. "Una rivendicazione: chi inventò il telefono?" *Il Corriere d'America*, February–March 1926.

Basini, C. "Antonio Meucci, pioniere dimenticato." *La Voce Repubblicana*, October 30, 1954.

Bayles, R. M. *History of Richmond County.* New York: L. E. Preston, 1887.

Bell, A. G. *The Deposition of A. G. Bell in the Suit Brought by the United States to Annul the Bell Patents*. Boston: American Bell Telephone Company, 1908.

Bianchi, Ugo. *La rivendicazione di una gloria italiana*. Rome: Tip. della Camera dei Deputati, 1923.

Bianucci, Piero. *Il telefono la tua voce*. Florence: Vallecchi, 1978.

Bocock, J. P. "The Romance of the Telephone." *Munsey's Magazine*, November 1900.

Bosi, Alfredo. *Cinquant'anni di vita italiana in America*. New York: Bagnasco Press, 1921.

Bruce, Robert V. *Alexander Graham Bell and the Conquest of Solitude*. Boston: Little, Brown, 1973.

Capelvenere, Franco. *Meucci, Storia di un'ingiustizia*. Rome: Newton Compton, 1989.

Capelvenere, Franco. *Meucci. L'uomo che ha inventato il telefono*. Florence: Vallecchi, 2003.

Casaretto, P. F. "Il ritorno di Garibaldi dal secondo esilio." *Nuova Antologia*, February 16, 1911.

Caso, Adolph. *They Too Made America Great—Lives of Italian Americans*. Boston: Branden Books, 1978.

Casson, H. N. *The History of the Telephone*, 9th ed. Chicago: McClurg & Co., 1917.

Catania, Basilio. *Antonio Meucci: L'inventore e il suo tempo*. Da Firenze a L'Avana, vol. I. Rome: Seat - Divisione STET, Editoria per la Comunicazione, 1994.

Catania, Basilio. *Antonio Meucci: L'inventore e il suo tempo. New York 1850–1871*, vol. II. Turin: Seat - Divisione STET, 1996.

Catania, Basilio. "Antonio Meucci Revisited." *Antenna, Newsletter of the Mercurians, in the Society for the History of Technology*, vol. 9, no. 1 (November 1996): 4.

Catania, Basilio. "Antonio Meucci: Telephone Pioneer." *Bulletin of Science, Technology and Society*, vol. 21, no. 1 (February 2001): 55–76.

Catania, Basilio. *Antonio Meucci: The Inventor and His Times*, 4 vols. Rome: Seat - Divisione STET, 1996.

Catania, Basilio. "Antonio Meucci. Una vita per la scienza e per l'Italia." Istituto Superiore Delle Comunicazioni e Delle Tecnologie Dell'informazione, April 2003.

Catania, Basilio. *The Meucci-Bell Dichotomy*. Milan: R&D Mond, 2001.

Cimagalli, Dino. "Il vecchio telefono entra nel futuro." *Gente*, 1984.

Coe, Lewis. *The Telephone and Its Several Inventors*. Jefferson, North Carolina: McFarland & Co., 1995.

Colonnetti, Gustavo. *Grandi primati italiani*. Alpignano (Turin): Alberto Tallone, 1961.

Conti, Giuseppe. *Firenze vecchia*. Florence: Bemporad, 1899.

Cucentrentoli, Giorgio. *Gli ultimi granduchi di Toscana*. Bologna: Edizioni La Perseveranza, 1975.

De Angelis, Marcello. *Le carte dell'impresario.*, Florence: Sansoni, 1982.

De Angelis, Marcello. *Le cifre del melodramma.*, Florence: La Nuova Italia, 1982.

De Land, F. *Dumb No Longer: Romance of the Telephone.* Washington, D.C.: Volta Bureau, 1908.

Desideri, Antonio. *Storia e storiografia*, vol. II. Florence: D'Anna, 1983.

Dwight, Theodore. *Electrical World.* New York: McGraw-Hill, September 6, 1884, and November 28, 1885.

Evenson, A. E. *The Telephone Patent Conspiracy of 1876: The Elisha Gray-Alexander Bell Controversy and Its Many Players.* Jefferson, North Carolina: McFarland & Co. Publishers, 2000.

Fantoni, Giorgio. *L'evoluzione del telefono. Dall'elettrofono di Meucci alla routine del cellulare.* Lecce: Youcanprint, 2020.

Favoino Di Giura, Giovanni. *Il vero inventore del telefono. Antonio Meucci.* New York: Cocce Press, 1940.

Ferrari, M. E. *Emigrazione e Colonie. Il giornale genovese La Borsa (1865¬–1894).* Genoa: Bozzi, 1983.

Ferrario, Artemio. *Invenzioni e inventori del XX secolo.* Milan: Bompiani, 1938.

Finocchiaro, Alfio. "Pronto, chi parla? Sono Meucci." *Il Messaggero*, Rome, March 22, 1986.

Gardner, Robert. *Electricity and Magnetism.* Minneapolis: Twenty-First Century Books, 1997.

Garibaldi, Giuseppe. *Le Memorie di Garibaldi.* Edizione Nazionale Degli Scritti di Giuseppe Garibaldi, vol. II. Bologna: Cappelli, 1932–1937.

Garibaldi, Giuseppe and Theodore Dwight. *The Life of General Garibaldi. Translated from His Private Papers with the History of His Splendid Exploits in Rome, Lombardy, Sicily, and*

Naples to the Present Time. New York: Derby & Jackson 1861; London: Forgotten Books, 2018.

Gay, H. N. "Il secondo esilio di Garibaldi, 1849–54." *Nuova Antologia*, June 16, 1910.

Ghibaudi, Bruno. "Il telefono." *Storia Illustrata.* Milan: Mondadori, 1968.

Gorel Porciatti, Umberto. *Simbologia massonica.* Rome: Anatòr, 1968.

Hamen, S. E. *Who Invented the Telephone?* Minneapolis: Lerner Publications, 2018.

Josephson, Matthew. *The Robber Barons.* New York: Harcourt, Brace and Co., 1934.

Langdon, W. C. "Myths of Telephone History." *Bell Telephone Quarterly*, vol. 12, no. 2 (April 1933).

Lattanzi, Agostino. *Antonio Meucci. Inventore del telefono.* Latina: Artegraf, 1998.

Latzke, Paul. *A Fight with an Octopus: Being the Story of a Great Contest That Was Won Against Tremendous Odds*, as printed originally in Success Magazine, Appendix H. February–June and August 1906.

Leng, Charles W. and William T. Davis. *Staten Island and Its People: A History, 1609–1929*, vol. 1. New York: Lewis Historical Publishing Co., 1930.

Lensi, Alfredo. *Napoleone a Firenze.* Florence: Rinascimento del Libro, 1936.

Lodge, O. "The History and Development of the Telephone." *Journal of the Institution of Electrical Engineers*, vol. 64 (1926): 1,098.

Lundrigan, Margaret and Tova Navarra. *Staten Island: Images of America.* Charleston, South Carolina: Arcadia Publishing; Lecce: Youcanprint, 1997.

MacKenzie, Catherine. *Alexander Graham Bell: The Man Who Contracted Space.* Boston and New York: Houghton Mifflin Company, 1928.

Maretzek, Max. *Sharps and Flats.* New York: American Music Publishing Company, 1890.

Marmier, Xavier. *Lettres sur l'Amérique.* Paris: A. Bertrand, 1928.

Menghini, Mario. *Lettere di Garibaldi, Q. Filopanti e A. Lemmi a Felice Foresti e lettere di Felice Foresti a G. Lamberti e a G. Mazzini.* Imola: Coop. 1909.

Meucci, Antonio. Caveat #3335 "Sound Telegraph." December 28, 1871.

Meucci, Antonio. Memorandum Book.

Meucci, Sandra. *Antonio and the Electric Scream: The Man Who Invented the Telephone.* Boston: Branden Books, 2010.

Mirola, Gian. "Antonio Meucci dette le ali alla parola." *La Nazione*, Florence, October 4, 1970.

Moncada, Francesco. "Antonio Meucci: Inventor of the Telephone." *Atlantica*, vol. 13, no. 5 (August 1932): 201–204.

Moncada, Francesco. "Incidents in Garibaldi's Life in America." *Atlantica*, October 1932.

Montanelli, Indro, and Marco Nozza. *Garibaldi.* Milan: Rizzoli, 1975.

Monti, A. "Un inventore amico di Garibaldi." *Le Vie d'Italia*, December 1936, 771–776.

National Italian American Foundation. *Italians in America: A Celebration*. New York: Mockingbird Press and Portfolio Press, 2001.

Nese, Marco, and Francesco Nicotra. *Antonio Meucci, 1808–1889*. Rome: Italy Magazine Publisher, 1989.

Nevins, Allan, and Henry Steele Commager. *Storia degli Stati Uniti*. Turin: Einaudi, 1978.

Newton, Roger. *Galileo's Pendulum*. Cambridge: Harvard University Press, 2004.

Niaudet, Alfred. *Téléphones et phonographes*. Paris: Baudry, 1878.

Paine, A. B. *Theodore N. Vail: A Biography*. New York: Harper & Bros., 1929.

Pearson, H. G. *Son of New England*. Boston: T. Todd Co., 1932.

Piccini, Giulio. *Memorie di un impresario fiorentino*. Florence: Loescher & Seeber, 1892.

Pizer, Russell A. *The Tangled Web of Patent #174465*. Bloomington, Indiana: AuthorHouse, 2009.

Prescott, G. B. *The Speaking Telephone*. New York: D. Appleton & Company, 1878.

Prescott, G. B. *Electricity and the Electric Telegraph*. New York: D. Appleton & Company, 1892.

"Profili di fisici italiani." *La rassegna di clinica, terapia e scienze affini*, anno XII, vol. IV (July–August 1942).

Reid, J. D. *The Telegraph in America and Morse Memorial*. New York: John Polhemus Publisher, 1886.

Respighi, Luigi. *Il telefono e la priorità di Antonio Meucci.* Rome: Edizioni Italiane, 1939.

Rhodes, F. L. *Beginnings of Telephony.* New York: Harper & Brothers, 1929.

Ridley, Jasper. *Garibaldi.* London: Constable, 1974; Milan: Mondadori, 1976.

Rossi, Adolfo. *Un Italiano in America (An Italian in America).* Milan: La Cisalpina, 1899.

Roversi, L. "Antonio Meucci a Staten Island." *La Follia,* March 15, 1953.

Savorgnan, Francesco. *Da Leonardo a Marconi.* Milan: Hoepli, 1930.

Schiavo, Giovanni E. *Antonio Meucci: Inventor of the Telephone.* New York: The Vigo Press, 1958.

Schiavo, Giovanni E. *Four Centuries of Italian American History.* New York: Vigo Press, 1952.

Schwartz, Ella. *Stolen Science: Thirteen Untold Stories of Scientists and Inventors Almost Written Out of History.* New York and London: Bloomsbury Children's Books, 2021.

Shulman, Seth. *The Telephone Gambit: Chasing Alexander Graham Bell's Secret.* New York: W. W. Norton & Co., 2008.

Snedden, Robert. *The History of Electricity: Advances That Have Changed the World.* New York: Thomson Learning, 1995.

Stein, Ralph. *Due secoli di invenzioni.* Milan: Mondadori, 1977.

Swezey, K. M. *Science Shows You How.* New York: McGraw Hill, 1964.

Taylor, L. W. "The Untold Story of the Telephone." *American Physics Teacher*, vol. 5, no. 6 (December 1937): 243–251.

Trevelyan, G. M. *Garibaldi and the Making of Italy*. London: Longmans, Green and Co., 1928.

Tuckerman, H. T. "Giuseppe Garibaldi." *North American Review*, January 1861, 15–56.

Tyrrell, H. "Garibaldi in New York." *The Century Magazine*, June 1907.

U.S. House of Representatives. H. Res. 269. 107th Cong., 2nd sess. (2002).

U.S. Supreme Court archives, Washington, D.C.: *Meucci's Deposition from December 7, 1885, to January 13, 1886*; *Deposition of September 11, 1886*; *Meucci's Affidavit of October 9, 1885*; *Judge Wallace's Decision, Federal Reporter, Vol. 32,729.*

Vannucci, Marcello. *Storia di Firenze*, vol. I. Rome: Newton Compton, 1986.

Virgilio, Jacopo. *Del commercio marittimo e della costruzione delle navi*. Genoa: Tipografia della Gazzetta dei Tribunali, 1858.

Watson, Thomas. *The Birth and Babyhood of the Telephone*. New York: American Telephone and Telegraph Co., 1942.

Watson, Thomas. *Exploring Life: The Autobiography of Thomas Watson*. New York: D. Appleton and Company, 1926.

About the Author

Cultural mediator, art curator, and translator Francesca Valente lives and works between North America and Italy. For over thirty years she brought the best of Italian culture to North America as director of the Italian Cultural Institutes of San Francisco, Toronto-Vancouver, Chicago, and Los Angeles. She organized many exhibitions of Italian art in the U.S. and Canada, such as *Leonardo's Return to Vinci*, which included four drapery studies by Leonardo da Vinci, and the *Salvator Mundi*. These precious artworks, owned by Giorgio Vasari, traveled to eleven university museums, from Berkeley to Princeton. While posted in Rome, she co-curated the collection of contemporary art of the Ministry of Foreign Affairs, La Farnesina.

Once back in North America, Valente was responsible for the permanent installation of Enzo Cucchi's *The Fountain of Italy* (1992) in Toronto and conceived and organized, among many others, the unprecedented exhibition *Leonardo da Vinci and Bill Viola* (2009) in Los Angeles. She coordinated the eight Italian Cultural Institutes of the U.S. and Canada, and curated permanent installations by Alighiero Boetti, Giò Pomodoro, Mauro Staccioli, and Eliseo Mattiacci.

In 2011 Valente was invited by the Italian Ministry of Foreign Affairs to coordinate the eighty-nine Italian Cultural Institutes throughout the world for the Padiglione Italia, Venice Biennale. In 2016–18 she curated for *Imago Mundi* an exhibition of 220 artists from Central and Eastern Canada featured at Palazzo Loredan, Venice, and at OCAD University Onsite Gallery, Toronto. In 2018–20 she curated three exhibitions of Nancy Genn's work (at Palazzo Ferro Fini, Venice; the Ca' Pesaro Museum, Venice; and Palazzo Collicola, Spoleto). In 2020–21 she was responsible for the first two editions of the *Festival delle Arti* in Todi, dedicated to Beverly Pepper, Gianfranco Gorgoni, and Arnaldo Pomodoro.

Valente recently published *A. P. Giannini: The People's Banker*, as well as *Rita Levi-Montalcini: Pioneer & Ambassador of Science* in bilingual edition for the Mentoris Project, Los Angeles.

Acknowledgments

I would like to express my gratitude to Jane Varnus and Filippo Mariano for helping co-edit my book. I extend my heartfelt thanks to Tania Moretto for her constant precious help, especially for coordinating the bibliography; and to Matilde Corda, Carla Plevano, Maria Lella Rebecchini, Helene Vukovich for their support of my project. As always, to Branko and Francesco, my husband and son, for being at my side from the very beginning.

Profound thanks are due to the Meucci Garibaldi Museum, Staten Island, in particular to William Castello and Marianne B. Prime for their assistance in my research.

Above all, I would like to thank Robert Barbera for his friendship and trust.

ALSO FROM THE MENTORIS PROJECT

America's Forgotten Founding Father
A Novel Based on the Life of Filippo Mazzei
by Rosanne Welch, PhD

A. P. Giannini—The People's Banker
by Francesca Valente

The Architect Who Changed Our World
A Novel Based on the Life of Andrea Palladio
by Pamela Winfrey

At Last
A Novel Based on the Life of Harry Warren
by Stacia Raymond

A Boxing Trainer's Journey
A Novel Based on the Life of Angelo Dundee
by Jonathan Brown

Breaking Barriers
A Novel Based on the Life of Laura Bassi
by Jule Selbo

Building Heaven's Ceiling
A Novel Based on the Life of Filippo Brunelleschi
by Joe Cline

Building Wealth
From Shoeshine Boy to Real Estate Magnate
by Robert Barbera

Building Wealth 101
How to Make Your Money Work for You
by Robert Barbera

Character is What Counts
A Novel Based on the Life of Vince Lombardi
by Jonathan Brown

Christopher Columbus: His Life and Discoveries
by Mario Di Giovanni

Dark Labyrinth
A Novel Based on the Life of Galileo Galilei
by Peter David Myers

Defying Danger
A Novel Based on the Life of Father Matteo Ricci
by Nicole Gregory

Desert Missionary
A Novel Based on the Life of Father Eusebio Kino
by Nicole Gregory

The Divine Proportions of Luca Pacioli
A Novel Based on the Life of Luca Pacioli
by W.A.W. Parker

The Dream of Life
A Novel Based on the Life of Federico Fellini
by Kate Fuglei

Dreams of Discovery
A Novel Based on the Life of the Explorer John Cabot
by Jule Selbo

The Embrace of Hope
A Novel Based on the Life of Frank Capra
by Kate Fuglei

The Faithful
A Novel Based on the Life of Giuseppe Verdi
by Collin Mitchell

Fermi's Gifts
A Novel Based on the Life of Enrico Fermi
by Kate Fuglei

First Among Equals
A Novel Based on the Life of Cosimo de' Medici
by Francesco Massaccesi

The Flesh and the Spirit
A Novel Based on the Life of St. Augustine of Hippo
by Sharon Reiser and Ali A. Smith

God's Messenger
A Novel Based on the Life of Mother Frances X. Cabrini
by Nicole Gregory

Grace Notes
A Novel Based on the Life of Henry Mancini
by Stacia Raymond

Guido's Guiding Hand
A Novel Based on the Life of Guido d'Arezzo
by Kingsley Day

Harvesting the American Dream
A Novel Based on the Life of Ernest Gallo
by Karen Richardson

How to Jump-Start Your Way to Real Estate Wealth
by Robert Barbera

Humble Servant of Truth
A Novel Based on the Life of Thomas Aquinas
by Margaret O'Reilly

The Judicious Use of Intangibles
A Novel Based on the Life of Pietro Belluschi
by W.A.W. Parker

Leonardo's Secret
A Novel Based on the Life of Leonardo da Vinci
by Peter David Myers

Little by Little We Won
A Novel Based on the Life of Angela Bambace
by Peg A. Lamphier, PhD

The Making of a Prince
A Novel Based on the Life of Niccolò Machiavelli
by Maurizio Marmorstein

A Man of Action Saving Liberty
A Novel Based on the Life of Giuseppe Garibaldi
by Rosanne Welch, PhD

Marconi and His Muses
A Novel Based on the Life of Guglielmo Marconi
by Pamela Winfrey

No Person Above the Law
A Novel Based on the Life of Judge John J. Sirica
by Cynthia Cooper

The Pirate Prince of Genoa
A Novel Based on the Life of Admiral Andrea Doria
by Maurizio Marmorstein

Relentless Visionary: Alessandro Volta
by Michael Berick

Retire and Refire
Financial Strategies for People of All Ages to Navigate Their Golden Years with Ease
by Robert Barbera

Ride Into the Sun
A Novel Based on the Life of Scipio Africanus
by Patric Verrone

Rita Levi-Montalcini
Pioneer & Ambassador of Science
by Francesca Valente

Roadmap to Financial Independence
by Robert Barbera

Saving the Republic
A Novel Based on the Life of Marcus Cicero
by Eric D. Martin

The Seven Senses of Italy
La Luna di Miele
by Nicole Gregory

Sinner, Servant, Saint
A Novel Based on the Life of St. Francis of Assisi
by Margaret O'Reilly

Soldier, Diplomat, Archaeologist
A Novel Based on the Bold Life of Louis Palma di Cesnola
by Peg A. Lamphier, PhD

The Soul of a Child
A Novel Based on the Life of Maria Montessori
by Kate Fuglei

What a Woman Can Do
A Novel Based on the Life of Artemisia Gentileschi
by Peg A. Lamphier, PhD

The Witch of Agnesi
A Novel Based on the Life of Maria Agnesi
by Collin Mitchell

For more information on these titles and
the Mentoris Project, please visit
www.mentorisproject.org

www.ingramcontent.com/pod-product-compliance
Lightning Source LLC
LaVergne TN
LVHW100140061126
851857LV00010B/166
9781947431591